THE SONIC THREAD
Sound as a Pathway to Spirituality

THE SONIC THREAD
Sound as a Pathway to Spirituality

By CYNTHIA SNODGRASS

PARAVIEW PRESS

NEW YORK

The Sonic Thread
Copyright © 2002 Cynthia Snodgrass
All rights reserved. No part of this book may be used or reproduced in any manner whatsoever without prior written permission except in the case of brief quotations embodied in critical articles or reviews. For information address Paraview Press, 1674 Broadway, New York, NY 10019, or visit our website at www.paraviewpress.com.

Book and cover design by smythtype.
Cover art by Shirley Harryman
Author photo by Patrick L. Scully

ISBN: 1-931044-37-6

Library of Congress Catalog Card Number: 2002107956

*To the True Teacher, the Holy Name,
the Divine Breath, the Great Silence,
the Creative Word.*

Contents

Introduction / 8

CHAPTER 1. *The Silver Clarinet: First Breaths* / 10

CHAPTER 2. *Sounds of the Wind: Early Practice* / 17

CHAPTER 3. *Music of the Spheres* / 27

CHAPTER 4. *Song Flight: Singing Lessons* / 41

CHAPTER 5. *The Word Made Flesh* / 54

CHAPTER 6. *Hospice Prayer* / 68

CHAPTER 7. *Being Still: "Don't You Miss the Music?"* / 84

CHAPTER 8. *Shamanic Journey* / 96

CHAPTER 9. *The Listening Lodge* / 109

CHAPTER 10. *David's Harp* / 124

CHAPTER 11. *Remembering the Names of God* / 139

CHAPTER 12. *Ancient and Sacred: Nothing but the Sound* / 153

CHAPTER 13. *Sonic Meditations: The Altar of Sound* / 168

Epilogue / 181

Bibliography and Further Reading / 189

Introduction

As far back as I can remember, I have been in love with the phenomenon of sound. Maybe it was because I was born "blind as a bat" and during my early years my ears learned to compensate for what my eyes lacked. As a childhood game I would close my eyes and pretend to be completely blind. I'd imagine that I could negotiate the world with only the use of sound to help me navigate through a maze of objects.

Maybe it was because I grew up with classical music in the house, accompanied by expositions on the importance of its meaning. Or, maybe it was because on hot summer nights, after dinner, my father would read out loud to the family long sections of ancient prose and poetry, old morals, and stories spun from some of the world's greatest sounds.

Or perhaps I was so fascinated by sound because there were so many wonderful things to listen to in the world–clicks and thumps, the whirrings of mechanical devices, the pounding rhythms and swirling eddies within the physical body, the exotic trills and glides of various bird calls–all so fascinating to focus on and to fathom as a child. There were times I even thought I could hear the sound of atoms humming in their orbits as their miniature worlds spun around.

Or maybe it was due to a dream experienced in adolescence, wherein I crouched, late at night, out of breath and in mortal fear, on the side steps of an Eastern temple as the sounds of a Dream-Wind saved my life.

Or it could be because every day for as long as I can remember, I have spent countless hours practicing some kind of musical instrument. As a child, each day I was required by my mother to practice intervals, repeat scales, memorize melodies, and count to the beat of a metronome.

Each instrument had its challenges–the single reed, the double reed, the bowed string, the plucked string. T.S. Eliot wrote that "between the ideal and the reality falls the shadow," and indeed, between my ideal of being able to issue forth radiant streams of celestial sound and the reality of God-given talent fell the shadow of hours and hours of practicing.

Intense listening in order to produce the right pitch, with fingers placed just so, was followed by the intricacies of coordinating left and right, high with low. The brain completed one chore while the body learned to automatically execute another. Rhythms were learned (three against two), chords were built from the bottom up, and melodies were spun from the center out. All for the sake of the beauty of sound.

It was beautiful, too. After hours of practice came the perfectly placed interval or the exact contour of a line that could open doors. Working in tandem or in chorus with others, harmony realized was a truly remarkable thing. When focused intently and breathing deeply, with the voice placed and tuned with consummate care, could come the amazing sensation of flying, the energetic explosion of having slipped the surly bonds of earth to glide and float high above the physical body. Music was a transcendental pathway and it could take one into higher realms.

Maybe it was in the fact that music has turned out to be my avocation, while a quest for the deeper perceptions of sound has continued to be central to my life's work, as well as to my spirituality. For, when I recognized vibration to be more than this world's music I realized how important the study of sound has been to my soul.

CHAPTER 1
THE SILVER CLARINET—First Breaths

I remember very clearly when I got my first musical instrument. I was in the fourth grade, and had been waiting a long time for music lessons to be available in my school.

Actually, I had longed to take singing lessons since the second grade. After school, while my mother was still at work, I would pretend to be a great opera diva. With grand arm gestures, I would take my theatrical stances in the living room or kitchen and proclaim the events of my day, articulated in the full operatic voice of a seven year old. The neighbors were probably less than grateful for my performances, but Mrs. Early, the second-grade music teacher who appeared once a month to teach music basics, had promised to take me on as a singing student when I was old enough. I looked forward with such zeal to that first singing lesson and cared very little for the neighbors' concerns. I sang, because I needed to be ready!

My first instrument was a silver clarinet. I will never forget that instrument. Although traded later for a wooden one that cost much more, there was something uniquely special about that first, fine instrument. A very close family friend had given it to me and it had been his when he was a child. So, the handing down of that instrument, in almost ritualistic fashion, given out of love and placed in my hands with a deep respect for music's depths, transferred to me an appreciation for the mysteries of music and an electric curiosity concerning the magic of sound.

In length, the silver clarinet measured just over half my own height, and at first it seemed that my fingers would never grow into the stretch that was required to reach its padded keys. As I removed the instrument from its case each day, the old box gave out its familiar creak and poured forth a smell that combined the perfume of aging leather with the mustiness of old fabric, its aroma speaking of the many years during which the silver had been surrounded in purple velvet and dampness.

In the daily ritual, I would take a moistened reed and carefully affix it to the mouthpiece, placing it evenly over the open space, and then turn the heads of the two silver screws that kept it in place. I connected the small mouthpiece onto the long silver tube, aligning everything in a straight line and carefully bringing the instrument into the correct position, perpendicular to my body.

In time, my gleeful curiosity and original excitement were often exchanged for questioning and frustration. I had a lot to learn on this instrument if it were to become anything less than a form of torture for all concerned! There was the matter of counting, the length of notes, the number of beats, the subdividing of beats. There was the matter of good pitch and pure tone. These resulted from correct finger placement added to the right mouth position, and sustained by sufficient breath support. My fingers had to stretch into positions that were unnatural; the muscles around my mouth grew weary from blowing and puckering. I grew dizzy as I squeezed the last bit of air out of my burning lungs just so the final tone of "Au Clair de la Lune" would not end disgracefully in a mighty squawk.

My music teacher was very patient. Each step of the way, he gave me exercises. Some were meant to strengthen my mouth muscles. Others increased the coordination and agility of my fingers. But the most important exercises were those for breath control, intended to improve tone quality and develop overall stamina.

The breathing exercises went like this. I took a deep gulp of air into my lungs and then played a note, releasing the air slowly until it was all gone, for as many counts as was comfortable. Then, I took in another breath of air, played the note again, releasing the air, but this time I increased the length of the note by one beat. This continued in like fashion, slowly increasing every day the length of time I could hold my breath, playing each note with a good, steady tone just a little big longer.

Once I could do this fairly well, the scales began, which were simply breathing exercises on different notes–up the scale and down the scale, down the scale and up.

At night, as I lay awake, reviewing the events of my day and waiting to go to sleep, I would hear the sounds of the silver clarinet echoing in my memory. The tones had shapes to them and some of them even had colors. I could sense the energy of the exercises still living in my awareness. I could see their colors in my mind's eye and feel their shapes resounding physically within me. And I practiced my breathing (or not breathing, as it seemed). I would hold my breath as long as I could, imagining I was underwater. Just when I thought that I couldn't take it any more, I would count out another five seconds, allowing the breath that I was holding to sustain yet another five beats.

This led to another exercise in which I concentrated on the shapes of the notes and began to expand or contract them. I could make each note as large as I wanted or as small as I wished. As I lay quietly with my eyes closed, at the end of the day, mentally practicing enlarging the sounds, I would focus my attention on the sound and visualize the memory of the note. Then I would enlarge the picture one step, and then another step. I would make the note as big as I could imagine. It couldn't get any bigger because it filled the whole universe! The color of the note washed over every existing thing. Its energy had expanded a thousand-fold. I would then take the picture of that largest note, vibrating in freeze-frame, and make it even bigger–and then even bigger than that. I would carry out this process an infinite number of times, and then at some point, I'd reverse it.

Slowly, the size of the note would return back to its

original shape, after which the process of reducing or decreasing it began, until it could hardly be seen, heard, or imagined at all. It was the smallest, barely perceivable dot, a tiny round circle of a note, suspended in time. Then, I would reduce the note ten times more, past a point which I could ever imagine, and then ten times more, and then ten times after that. With each of the reductions came the sensation of a pulsating energy, like light, being squeezed from the form each time it was compacted. At some undetermined point, somewhere in the unseeable darkness, I would again reverse that process, bringing the note back into my awareness and back into the realm of hearing.

When doing these exercises, a strange sensation would occur and gradually take over my awareness. It was as if, with the expansion and contraction, I could feel the world itself breathing. It felt like I was floating and I could feel how the world was actively alive. Ordinary objects in my room, any of which could be expanded or contracted in like fashion, were alive and pulsing with energy. So were the animals, and so were the cars. And so was I. I was connected, in those same expanding and contracting forms, to the rest of the world, to the rest of the universe. That was the first time I remember being aware of that sensation and of feeling "at one" with the great sea of vibration I sensed all around me. I was somehow connected to a vibrating presence much larger than myself, and yet part of myself. The realization of this connection brought me a great deal of peace and a sense of well-being.

As I lay awake in the dark, recalling the instrumental sounds that had been made during the light of day, it was as if I were remembering sounds made in some distant place that I knew well but with which I was no longer in touch. It was as if a very fine veil had been lifted by the exercises and I was encountering deeper memories of sound within. One set of sounds had been given to me by my music teacher; the other I knew internally, somehow, by heart. It was as if the one had awakened echoes of the other.

At the time, I simply enjoyed the exercises, the interplay of sounds, and the peace found within.

I didn't know it then, but those exercises were the prelude to a life-long study of music, devotion, and sacred sound. With that silver clarinet, the first of my magical music lessons had begun.

This story, and each of the stories to come, is followed by a section of reflective musicological notes which are intended to give the reader more information about a particular topic or tradition. A bibliography for further reading (as referenced in the notes) is provided for each chapter at the back of the book.

This first chapter began by weaving together several vivid memories from my childhood about sound and spirituality. First is the memory that I was never happier than when singing. Second, I remember spending hours lost in activities that were meditative in nature, many of which involved the use of the breath. The gift of my first musical instrument enabled me to combine those two joys—the creation of sound and meditation—through the art of learning how to listen.

Young children are innately musical; they make up songs freely, singing contentedly to themselves or others for hours. For children, "[m]aking up songs is an inclination as natural as breathing and as natural as our heart-beat" (Hale, 64). However, at roughly age four, children lose their contact with spontaneous singing in order to focus on "the learned song" taught to them by adults (Campbell in Hale, *ibid.*). Gradually, with the onset of a larger spoken vocabulary, children lose the musical sounds in their speech, and, eventually, with the direction to "say it like a grown-up," the message of not using spontaneous musical expression becomes fully integrated for most children, thus quieting the natural and innate musical impulses, often for the rest of their lives (Katsh and Merle-Fishman in Hale, *ibid.*).

Many children also have a natural affinity for meditation or contemplation. Unable to take it for granted, I was fascinated by the breath. Breath is life itself, and as such is directly spiritual. Our words "spirit" and "spiritual" are rooted in the Latin *spirare*, meaning "to breathe" or "to blow." Likewise, in Greek, *pneuma* can mean either "breath" or "soul," and *psyche* can refer to "soul" or "mind" or "a breath," or "a gust of wind." The Sanskrit term *aatman* can be translated "soul" or "self," "air" or "wind," just as the Latin word *anima* is used interchangeably to mean "soul," "air," or "breath." We are animated, made soulful, in-spirited, through the breath.

Breath, or spirit, is the very enervating stuff of life. In the book of Genesis, it was *ruach*, that great wind of spirit, that moved across the face of the Deep at the time of creation. It was breath that the Lord blew into the lifeless clay destined to become humanity. In certain parts of the early Church, the bishop's breath, blown into a pouch, was carried to faraway places for the ceremonies of reception into the community of faith, the breath being released from the pouch onto the new initiates as a symbol of new life and new spirit. In India, it is called *praana* or *shakti*, and in China, it is known as *ch'i*. All of the world religions, from Judaism, Christianity, and Islam, to Hinduism and Buddhism, as well as the indigenous religions, revere the breath as sacred.

To be aware of the breath is to be aware of the sacred. Breath is the infinite self; the very act of breathing–inhalation and exhalation–is doing what God is doing in the process of creating everything that we know on this plane. "We [are] made of appearing and disappearing light that came from the inhalation and exhalation of God's breath" (Rael, 23). Further, scientist Itzhak Bentov has found that during the pause in the breath, during the still point that comes between the out-breath and the in-breath, the body itself produces a wave form that oscillates at around 7.8 cycles per second. "This is a frequency which is believed to be the resonant frequency of the Earth, and that during this time, we are, however briefly, also locked in resonance with the energy of the Earth" (Goldman, 139). In the ancient practice of *ch'i kung*, "embryonic breathing" was cultivated so that within the physical body a subtle body was gradually born, and "it was through this [subtle] body that one achieved immortality" (Kramer, 87). To be aware of the breath is to be aware of the sacred, and awareness of the breath is fundamental to good musicianship.

Breathing for the musician then is (at its core) a spiritual act. Another spiritual act, in which anyone singing or playing an instrument is continually engaged, is that of listening. Normally, we listen to hear what is being said or to communicate. But good listening goes much deeper.

We listen in order to be in tune—in tune physically, in tune mentally, in tune spiritually. Musicians listen continuously to themselves, as well as to those with whom they are playing, to check whether they are all in tune. The more we hear, the more we are able to adjust and align ourselves, the more we are able to come into harmony with our selves and with others. But in order to be in tune we must develop the true ability to listen. "The true listener is no longer defined by desires or attachments. Instead, he or she is sensitized to consciousness" (Rael, 67).

Real listening takes one beyond the limits of self. "At best, true listening is an approximation of selflessness whereby one person opens not only their ears, but their heart to the words, spoken and non-spoken, of another" (Weeks, 48).

CHAPTER 2
SOUNDS OF THE WIND—Early Practice

Some people's talents are clear and self-evident from the very beginning. My talents, however, started off rather hazy and have developed slowly over the course of my life.

As already mentioned, one of my loves has been the study of music and sound. Another has been the abiding awareness of the presence of something divine in my life and in the world around me. In my experience, the "unfolding" that has been my natural life seems to have been about unraveling and weaving together those two threads.

I have always been aware of being very close to something divine—an entity, an energy, or a presence that has been called variously our higher good, the Great Spirit, the Ground of our Being, or just plain "God." But this closeness, this sense of personal relationship, hasn't always come in the arena of organized religion. As a child, I don't necessarily remember feeling closest to that Presence when I was inside a church.

Although I was born to an allegedly long line of Presbyterian ministers on one side of my family, I was raised Unitarian. This was not a choice I made consciously, but a parental decision made on my behalf. At the age of five, I remember walking down the aisle of the Unitarian church to receive a little potted marigold plant, the sign of being welcomed into the realm of growing things. Our hymnal had special hymns in it, the verses of which praised Moses, Mohammed, maybe Jesus, and Thomas Jefferson all in one song. The Sunday school curriculum included classes on utopia, physical science, civil rights, and community building.

Our church was the only church in town that did not have a cross on the top of it—it had a lightning rod, which actually served a very practical purpose! Somehow, I identified with that lightning rod. While an in-joke for those in the church, it was also a powerful symbol, one that signified readiness and the ability to ward off attack. Children who belonged to *that* church needed to ward off attack. Kids can sometimes be cruel to each other, but when it comes to religion I have a suspicion that they can be particularly vicious. The taunting and needling that I received, delivered ostensibly in the name of Jesus, were enough to convince me *definitively* that I wanted nothing to do with a religion whose followers were so mean-spirited and deliberately cruel.

Regardless of church affiliation or its associations, I spent a good deal of time in early adolescence pondering the workings of the universe, as well as the meaning or purpose of life. I was fascinated by the myriad of possibilities within creation that might explain what on earth was going on in this so-called "existence" and why. My favorite possible answer, at the time, was that the world was a dream within a dream within a dream–the visualizing of which could provide fascination for long periods of time. The idea that the universe could wake up and fall asleep an infinite number of times was for me intriguingly seductive.

My dreams were also very vivid during that period, and that may have had something to do with why I delivered a speech to my eighth-grade class on that very topic. I informed the class that we were not truly as we seemed, because one of us was not in the classroom at all, but was *really* home in bed dreaming all of us up! We could never be sure who that person was, but the classroom did not truly exist outside the dreaming mind of that one student. I explained that the classroom did not exist and the desks did not exist. Neither the students nor the teacher existed. We were not really *real*. However, the message was not a matter of bleak existentialism. For, while the dream scenario was true, and we couldn't do anything about that, the *real* message of good news was that we were, indeed, all in it together! No doubt the other students considered that if it wasn't *me* home in bed dreaming all this up, then it was *surely* me out to lunch!

Just the year before, I had begun to study a new instrument–the oboe. The oboe was a fascinating instrument. It was exotic and had a very unusual, reedy tonal quality. The music written for the oboe reminded me of faraway, foreign places, like the Middle East, Mongolia, or even India. Images of snake-charmers came to mind as I would envision a thin, dark-skinned man wrapped only in a turban and loin cloth sitting in front of a tattered wicker basket, hypnotically weaving his magically reeded instrument back and forth before a wide-hooded, poisonous snake, whose body wound its way up and out of the basket, tongue flickering, its glowing eyes glued lovingly to the invisible notes in the air.

Oboe players automatically played in the school orchestra as well as in the school band. All *other* wind instruments had to try out for the few seats available in the orchestra. That was wonderful in and of itself, because playing in the orchestra gave me access to a whole range of orchestral music, "real music," that was never attempted by the junior high band.

I took to the instrument right away. Its body was smaller and it fit in my hands better than the clarinet had, and I didn't have any trouble getting sound out of the small double reeds. I loved its sound. The reedy quality had a way of cutting through the air that no other sound had. It was sharp and attentive, yet plaintive and sad. It was thin, and yet full. The sound was expressive and fun to play with. The music teacher told me that I was a "natural" on this instrument, and I seemed to progress very quickly. It was as if I had made those sounds somewhere before.

For me, being a "natural" soon came to mean that I was personally relieved of the need to practice. I became completely undisciplined, but I was still able to learn things on that instrument so quickly that weekly lessons seemed a breeze. I might practice once or twice a week before a lesson, but not all the time. More likely than not, I would learn the lesson about an hour or so before my actual lesson time. I would play through the exercises and review the little melodies several times, making sure I had covered everything. Then I would give myself fifteen minutes or so for my mouth to recover so I'd have the muscle tone to actually play through the lesson. There were weeks, no doubt, when my music teacher was fully aware of this short-cut approach to developing technique, but there were many weeks when it seemed he didn't even notice. He was pleased, overall, with my progress.

I also found that reading through my music and inwardly rehearsing each note and physical movement was almost as good as actually practicing. One day, after lunch, I was sitting in study hall reading through my music. I had physically practiced the night before, but my lesson that day was scheduled for later in the afternoon, and I wanted to make sure that I read through the pieces one more time before having to actually play them. It was early in the summer on an unseasonably hot and muggy day, and although the windows in the study hall had been opened, they weren't very effective in granting relief from the heat. It was stiflingly hot and lunch had been heavy. My eyelids drooped slowly closed. They would snap open, and then droop again. The notes of music swam hazily in front of my dazed field of vision.

I heard the melodies in my mind; the very clear and nasal tones of the oboe filled my being. Gradually, I began to hear the sounds of something outside of the oboe's tones that sounded like insects. The sounds began to whir around me, their tinny rattles buzzing at me with sounds that changed timbre, from copper to silver, from silver to gold. The buzzing became a humming, and the humming became another kind of clear but definite tone. Along with the tone came the sensation that I was riding on top of it. The very top of the sound was moving, yet it was somehow solid, and it was possible to sit on top of the tone, gliding along the top of it as it moved, allowing the swells of the tone to carry me.

The swells were like waves, but the waves weren't exactly like water. They were more like mist or like sea foam, or maybe dark smoke. They kept me from seeing where I was going. I was simply traveling, moving along the tone in a swirl of mist and darkness. Eventually, after the feeling that some distance had been covered, and even that worlds had been crossed, I was able to open my eyes. I saw a woman. She appeared to be from India. It was still very dark, but I could see that she was working, or rather serving, in a temple. As I gazed into her eyes, I felt myself blending into her; her mind was becoming my mind, and as we merged I could tell that we really were no different, she and I.

I could feel that she was breathing heavily. *I* was breathing heavily too. She crouched, out of breath. *I, too*, crouched, out of breath, some time late in the night, hidden on the side steps of an Eastern temple, in mortal fear for my life. I was fixed in time, afraid to move for fear of being found clinging to the steps that led to the main altar, clutching the cold stone, hoping that the shock of the frigid, flat coldness might awaken me or bring me back to myself.

In the dark air held within the structure's mossy stones hung a dank mist made partially of fog and partially of smoke. Sweeping currents of sweet black fragrance were mixed with the flavors of pungent incense and the fruits of offering, while mists moved in the darkness, covering the black silence in heavy, velvet curtains. A hush, echoing in the fading memory of recent activity, held the sacred space of the temple in expectant quiet, with only the sound of my breathing to break the silent air.

A low wind began to moan past the arches of the doorways that punctuated the walls of the temple. The sound of that wind began to grow and to reach toward me; wisps of whirling smoke spun itself around me, enfolding me. The clouds of mist encircled like a cocoon, protecting me from sight. From this vantage point of cloud, I watched as two men with flailing sabers raced past my hiding place, fooled by the sound of the whirling wind. As they hurried past into the outer rooms of the temple, the whistling gradually began to subside. The intensity gradually decreased. Eventually, freedom was mine, and I stealthily followed the last whispering wind-sounds through the doorway and out into the covered courtyard hidden by the night.

I was breathing deeply as the buzzing subsided and the sounds of the oboe disappeared from my ears and brought me back to the heat of the study hall. To this day, I do not know whether what happened was real or unreal, dream or vision, but I know that it was the first time I came to know sound as a method of travel and that music could take me places!

After class, I went to my lesson, just like any other week. But that day, after the rudimentary pieces had been played, both my teacher and I remarked on having had a strange, but similar, sensation. We both had noticed that during the lesson it seemed as if the walls of the small, stuffy practice room, which was tucked beneath an ancient, creaking staircase, had somehow released from the depth of its sound-proof tiles the slightest hint of that fragrant smell that incense brings.

Just as the sounds of the breath are sacred in many cultures, a special respect is also given to the wind and its sounds, which are referenced in this chapter. The winds of the four directions are viewed as sacred, as they frame and energize the world. The wind signifies life itself. *Ruach*, the spirit-wind, hovered over the Deep at the beginning of creation. In fairytales told worldwide, death comes and goes on the wind, and thus developed the custom of leaving a window open during the time of one's dying, as it is believed that the soul leaves the room on the sound of the wind.

While holy wind is connected with the beginning and the end of life, it can also act as a guide or spiritual counselor during life. In the Jewish mystical tradition, the "soul-breath that enters at birth directs and trains the human being and initiates him into every straight path" (Abram, 247), being a kind of force that guides one and teaches. In Native American traditions, "smaller winds" grant powers of perception to the individual. The smaller winds are ways of knowing, ways of being given information. "These two Little Winds linger within the spiraling folds of our two ears, and it is from there that they offer guidance to us, alerting us to near and distant difficulties, helping us to plan and to make choices" (*ibid.*, 234).

The wind is connected with worship and inspiration. Forms of initiation may come through the wind and divine information may be granted through its sounds. In the *Rig Veda* (X:70.8) of ancient India, the Spirit-wind is identified with the blissful release of *soma* that comes in the perception of "the raw moment of living . . . [which is] the true guide, the original guru and world-teacher . . . [whose] spontaneous release is what frees us, not our effort to be free or our thoughts of freedom, which by their fixation inhibit the flow" (Frawley, 195). It is described in the rituals of India as the *pavan* of adoration, which can possess the body during worship, catching up devotees, causing them to dance or sing, lost in the ecstasies of divine presence (Erndl, 107). It is described in the New Testament as a sound "from heaven like the rush of a mighty wind," which filled the house where Christ's disciples had gathered at Pentecost, resting upon them in "tongues as of fire" (Acts 2:2).

Finally, the four sacred winds also enable us to speak and to communicate. For the Navajo, "the most potent force in the universe is wind, given form in speech and song" (McAllester, 52-53).

Other aspects of music deepen our awareness of the sacred as well. The learning of any instrument, like the exploration of one's voice, is a spiritual journey whose goal is defined not so much by some obvious mastery proclaimed at the end of the road, but rather through the practice of lovingly repeated details.

Music is full of repetition and filled with practice. As a musician carefully and lovingly gets to know his voice and/or instrument, a kind of rock-polishing process takes place and the long hours of practice turn into a deep friendship. It is a friendship of patience, one that is not in a hurry, for "[h]aste has the potential of turning one's efforts into mechanical achievement rather than authentic evolvement. The process should be allowed to evolve at its own pace through patience and consistent practice" (David, 99).

The repetition of exercises establishes a point of focus from which deeper insight can evolve, perhaps even an altered state of consciousness. "To some scholars, a short musical phrase repeated and repeated and repeated (and repeated) has an intoxicating effect," (Reck in Gardner, 57), and repetition is a large part of all types of music that use sound to induce trance or altered states.

Repetition also develops mindfulness, as each musical exercise demands that it receive the full focus of one's attention. "It is the steady work in which one gently, firmly, and consistently brings oneself back to the task at hand that [. . .] facilitates the personality reorganization that is part of our slow, endless growth to real maturity" (LeShan in Roth, 72). Mindful repetition develops one's ability to focus one's attention, whether the musical exercise happens to be listening (in order to be in tune), or counting, or observing how the hand is placed just so in a certain fingering position. When practicing the individual elements of music, one puts all other distractions out of mind, and just as in meditation, the aim is "to fully occupy the present moment, free from fears arising from past experiences or concerns for the future; to control the reactions of [one's] ego; [and] to move out from under the burden of constant self-judgment" (Houston in Ferry, 1).

From the mindfulness of repetition comes a one-pointed concentration. Music trains "the mind and the soul, for music is the best way of concentration [. . .] keeping all other things away, one naturally develops the power of concentration" (Khan, *Music of Life*, 34). This brings with it clarity and understanding, with which "we can turn our attention to any problem and penetrate to the heart of it" (Easwaran, 213). Thus, the power of concentration found in the study of music translates to other areas of our lives.

Another spiritual connection for the music lover is found in the complexities of rhythm. Rhythm has to do with time and with movement, with counting and with accent. Ultimately, rhythm has to do with progressive order. In ancient India the term *rta* meant order or law of the universe, especially as being the creator and sustainer of the world. (It is possible that we still hear the Sanskrit word *rta* in our English word "rhythm.")

Rhythm is the first impulse, which proceeds from the Unmanifest Silence into the manifested world in the form of Time, and as such it is a bridge between the spiritual and the material worlds. Likewise, the human body is comprised of rhythm, from the walking gait of the mother perceived by the baby in the womb, to the steadiness of the heartbeat, to the electrical impulses within the brain. As the musician engages in rhythmic activity, he or she reconnects his or her own body with the manifesting process of cosmic order. "Through music making, one sets out on the path that leads to participation (or, as they say, 'drinking') at the river of cosmic rhythm" (Schneider, 75).

Steven Halpern has written that "music and the presence of God have long been associated," and that "in all cultures, people have used music to facilitate meditation and religious experience" (Halpern, 181). Music is a transcendent medium, capable of creating altered states of consciousness, able to connect us deeply with the primary forces of creation, regardless of the language we use to name them. By concentrating on the elements of music, one is brought into direct contact with those sacred forces that underlie all of creation.

CHAPTER 3
MUSIC OF THE SPHERES

I spent my undergraduate years in the Renaissance. They were woven from a rich and colorful tapestry of studies that ranged from language and philosophy to music and the arts. Fascinated by the currents of cultural rebirth, I immersed myself in subject after subject, searching for the essence of the era and hoping to encounter the spirit of the age.

A study of the Tempio Malatestiano at Rimini, with its *bas relief* musical figures and its astrological symbols and arcane references, had resulted in a marvelous journey illuminating the everyday lives of the great *conditierri* families of the Italian Renaissance, the rise of the city-states, and the birth of the Florentine Academy. However, this journey had left my mind swimming in a sea of images whose deeper meanings and inner-connectedness could only be deciphered through an in-depth investigation into the philosophical and religious beliefs of the time.

After standing on a course registration line for hours, I was advised by a graduate student sitting behind the desk that in order to take an upper-level course such as Italian Renaissance Philosophy, I would first be required to take a lower-level course offered by the department, Logic. So, trustingly, I signed up for the lower-level course, postponing the true investigation until the simple but requisite preliminary was fulfilled.

As it turned out, the investigation was postponed for some time, due to the fact that I ended up having to take Logic twice! It just was not within my mental vocabulary or ability to distill the fullness of real events into a series of seemingly artificial statements ($a = b$; $b = c$) that resulted in an equally neat final equation ($a = c$) that left everything all neatly tied up in a bow. It was a gut-level reaction. I objected to the very concept of Logic at a primal, cellular level. Consequently, I was mortified when I finally had to admit defeat, taking an incomplete the first semester, in order to return to it again a second semester with the help of a tutor, only to muddle through it again, gaining the hard-won but embarrassing grade of C!

It goes without saying that I was struck completely speechless when, weeks later, in relating this sad saga to one of the upper-level philosophy professors, he exclaimed: "Logic? You don't need the concepts of modern Logic to count the number of angels who can dance on the head of a pin!" I continued to explain my original desire to investigate the arts of the Renaissance, especially music, and to understand how the philosophy of the age had given birth to that creative spirit. My mood brightened as he replied, "Yes, it's a very intriguing subject! However, it is very much like an old, medieval city–many, many ancient and winding streets, with lots of little alleyways–very easy to get lost there! You will need a guide!"

Exhausted from the long battle with Logic's theorems, and elated by the good fortune of having finally found someone who understood what I was saying, I gladly and willingly followed my "guide" down one winding street after another into the heart of the Renaissance.

The subject was maze-like, as it encompassed not only philosophers and thinkers from one era, but also those classical and post-classical philosophers whose works had been re-discovered and assimilated into fifteenth- and sixteenth-century thought. The Renaissance mind, engaged in breaking from the confines of medieval attitude and perspective, had found in the human being the pinnacle of God's creation, a creature who stood just below the angels in ability and creativity. And, in order to grow beyond the boundaries set by medieval Church and society, the philosophers of the Renaissance reached back into history through newly discovered and recently translated manuscripts to grasp again the wisdom of the Ancients.

My "guide" promised that we would traverse both the major thoroughfares marking the tensions between the revealed insights of the Christian religion and the natural wisdom of *sapientia* proclaimed by the Ancients, as well as some of the smaller streets, which would also satisfy my quest for things musical. In reality, music was never far away, as it had been an integral part of ancient cosmology, and it was still deeply rooted in Renaissance philosophy as well.

One of the major thoroughfares of the journey was the Florentine Academy. Cosimo de' Medici, an affluent Florentine banker fascinated with ancient theologians such as Plato, Pythagorus, Orpheus, Hermes Trismegistus, and Zoroaster, as well as Jesus Christ and Moses, supported the Academy in recovering and translating ancient Greek manuscripts that came to shape the Italian Renaissance mind. Cosimo's primary translator was a priest named Marsilio Ficino, and as head of the Platonic Academy in Florence, he translated all the works of Plato, the later *Corpus Hermeticum* of Hermes Trismegistus, and many of the Neo-Platonists into Latin.

Music held a significant place in Plato's designs for society, since music was thought to have a profound effect on people's characters. Moreover, music shaped a large part of his cosmology, as it was through a musical process that individual souls came in and out of existence. In "The Myth of Er," Plato described the planetary universe as consisting of layers, like a stack of eight mixing bowls, with a siren-muse-type figure standing atop each rim. Each muse turned her bowl by its rim, each bowl emitting one note, bringing the total number of notes to eight–an octave. Above the bowls were the thrones of the three Fates, each of whom sang to the music emitted by the spinning spheres below. Lachesis sang of past events; Clotho sang of current events; and Atropos sang of things to come. Individual souls, then, migrating from one life to the next, made a circuit in front of one Fate, then the next, and then the next, threading their way in a musical process back into the world to live out whatever lot in life they had just chosen.

This understanding of the "Music of the Spheres" was elaborated on by many later generations. In his work "The Pimander," Hermes Trismegistus described the rising of the soul at the time of death, which involved releasing the material parts of the body and surrendering to each of the heavenly spheres in a purification process that brought the soul into a rarified state and then into the presence and light of the Divine Mind.

In *The Dream of Scipio*, Cicero wrote about the ascent of the soul through the planetary spheres. Each sphere issued a distinct tone. These tones were the "keys" to the universe. While most mortals could not hear the planetary tones, those who could imitate them with stringed instruments or through singing were granted access to the higher realms. As late as the time of Robert Fludd, in seventeenth-century England, it was believed that the musical intervals themselves (the octave, the fifth, the third) served as connectors, or gateways, between the music of the outer spheres and the music of the inner spheres. Simply by practicing and studying the intervals one could attune oneself to the higher realms.

Marsilio Ficino, a physician and priest, went to great lengths in reviving the Platonic and Neo-Platonic philosophies and harmonizing them with the Christian creed. Drawing on the "Music of the Spheres," Ficino referred to the sounds made by the planets as they moved in their orbits as *musica mundana*. The harmonies and rhythms found within the human body were known as *musica humana*, and the physical sounds issued by the human voice or a musical instrument were *musica instrumentalis*. It was believed that if one properly aligned oneself with the harmonies of the universe (*musica mundana*) then the inner harmonies of one's physical and emotional being (*musica humana*) would become balanced, allowing inner concord to be manifest. This alignment was obtained, of course, by using either the voice or a musical instrument (*musica instrumentalis*). Ficino himself used both the singing of celestial songs, while accompanying himself on the *lyra da braccia*, to attract the influences of the planets and align his own *musica humana* with the universal *musica mundana*.

Ficino also prescribed a more "orthodox" use of prayer. By asking for wisdom from Wisdom, it was possible to experience, through pure contemplation, a divine light flooding into the mind, followed by mystical union with that Light. This *summum contemplationis fastigium* had been referred to by Plato as "beatific vision," and Pico della Mirandola had identified it with the visionary sight of Moses, Paul, and St. John as they came face to face with divinity.

Pico della Mirandola was another philosopher of the Florentine Academy. Originally a protégé of Ficino, he traveled to Rome in 1486 with his nine hundred theses, the *Oration on the Dignity of Man*, in which he attempted to demonstrate that the Hebrew, Greek, and Christian traditions were all compatible, being joined together by one fundamental, mystical strand of knowledge. Barely escaping complete censorship from the Roman religious authorities, Pico practiced the *Kabbalah*, an ancient source of Hebrew wisdom.

Though primarily developed in medieval Spain, the *Kabbalah* was attributed to Moses as secret knowledge that he had handed down orally to certain initiates and which explained mysteries not written down in the book of Genesis. The *Kabbalah* was based on the *Sephiroth*, or the ten names of God, and it concentrated on the mystical properties of the Hebrew letters and words. By calling on the names of God, the *Kabalist* could enter into a state of deep trance, coming eventually into deep union with the divine. With a true understanding of the letters, a *Kabalist* singing a Psalm could come to a place similar to that of Hermes Trismegistus in "the Pimander," wherein there was a Receiving (*Kabbalah*) of the song of transcendental knowledge. This communion was so profoundly intimate and divinely ecstatic that it was possible for the soul, lost in divine harmony, to separate completely from the body, causing the person to die in a state of such indescribable bliss that is was known as *mors osculi*, or the Death of the Kiss.

These, then, were but a few of the ancient and fascinating metaphysical inquiries that we traversed, one after another, as my "guide" led me down one street and into another, through this alleyway and into that, in our quest to understand the philosophic thread of Renaissance awakening.

However, another very important thread that helped create the rich subtleties of texture and the depth of color of that Renaissance tapestry was my study of a stringed instrument from the period, the *viola da gamba*. I was in love with that instrument from the moment I first heard it. Its sound had an otherworldly quality, the kind of sound that evoked images of light. It was, unfortunately, that same lightness of tone that was the cause of its eventual demise.

The *viola da gamba*, a stringed instrument held between the knees and played somewhat like a cello, could not produce the magnitude of sound needed to sustain itself in the larger public auditoriums that came into vogue with the eighteenth century, and so the instrument was eventually disregarded. However, the lighter tonal quality that had been perfect for chamber music and the more intimate settings of the Renaissance had captured my imagination. My greatest desire was to learn how to make that instrument sing!

I studied with a teacher from a nearby university, and his teaching methods were meticulous and quite precise. Slowly, gracefully, one moved the left hand from one position to the next. Placing the hand in the correct position, one then watched carefully as it moved, precisely in slow motion, into the next position of perfect placement. My teacher could analyze any gesture and break it down into a series of isolated movements, repeated over and over again, until they were perfectly embedded in the muscle memory.

The right hand held the bow, fingertips grasping the bow-hair at one end while connecting with the strings of the instrument at the other. One had the very tactile sensation of actually pulling sound out of the instrument with the right hand. Each string had to be sounded in just the right spot. Hours of practice were required to perfect one note. The bow, moving at just the right speed, had to cross the string at just the right angle and then switch to the next string at just the right spot, and thus to the next string in an intricate complex of geometric progression. The arc of perfected bowing stretched from top string to bottom, vaguely echoing the outline of the bridge of the instrument, and then back up again. I spent hours simply repeating the pattern of this arc, rocking the bow silently from one string to the next without making a sound.

The process of creating sound on the *viola da gamba* was like patting the head while rubbing the stomach; two very different kinds of gesture were involved. The left hand executed one series of exact movements, while the right hand performed another, and the two were combined together to produce the ultimate result of music. My teacher referred to the whole process as "crying with one eye." One had to isolate and control the individual movements of both sides of the body to such a high degree that it would be like gaining control over both the inner mind as well as one's outer physical movements, so one could allow tears to be released from one eye, while directing those of the other to trickle down somewhere deep inside. I sat for hours, working to perfect the art of "crying with one eye."

The music written for this instrument took me to many fascinating places. I could travel from the great festivals of the Medici in Florence, to small chamber gatherings in Tudor England, to the courts of Henry VIII or Louis XIV–even to the grand and opulent fetes held in the Galerie des Glaces at Versailles. I could attend a seventeenth-century French gall bladder operation or attend church services in Venice. I could walk with Jesus, tread his tragic steps up the hill to Golgotha. Christ is portrayed by the *viola da gamba* in Bach's "St. Matthew Passion." But before I could begin to tackle the repertoire, I had to spend endless hours on the practice of "crying with one eye."

And so it was that I sat, one late afternoon in spring, in a small apartment of an old Victorian house, just at dusk as the sun was melting in the sky. I stared periodically out a large leaded-glass window as my awareness shifted back and forth between the mental images of the philosophy I had been reading and the physical shapes that my hands were making as I played the viol. I watched as my fingers placed themselves on the strings, making contact at just the right spot, ensuring that the interval would be pure and clean, moving then to the next position where two notes would entwine, aligning themselves in perfect harmony.

Surrounded by the transcendental cosmology of Plato, the musical gateways of Ficino, and the mystical bliss of Pico, I was at that moment reflecting on the thoughts of a sixteenth-century French philosopher, Bovillus, who taught that "wisdom" was the knowledge that one immaterial substance has of another immaterial substance. Suddenly, I noticed that the light coming through the window was beginning to grow brighter.

The light, which had been fading, began to intensify. It started to shimmer and then to dance. I was looking intently out the window to the vanishing point, where it seemed that something was forming, when I perceived that the image was a bird. A bird was gradually making its way towards me, a beautiful white bird. Its feathers were soft and its wings moved gracefully and effortlessly towards me.

It caught my gaze as I sat transfixed in awe.

The bird gazed directly into my eyes, and as it did so the feathered wings, arched upwards in mid-flight, were transformed slowly, as if changing from liquid into solid form, into the white wimples of a religious habit. The feathers disappeared and I found myself face to face with the most beautiful woman I had ever seen. Her features were delicate and graceful, as if crafted from fine porcelain. Her eyes expressed a deep kindness as she enfolded me completely within the compass of her loving gaze.

Just as the feathers had disappeared, so did the white wimples. They vanished into the air behind and above her, and I was left staring at this beautiful, vibrant being. She was wrapped in sheer, luminous cloth that gave off different colors, and with each movement she radiated harmonies that were left hanging in the air as she moved. Around her head circulated a small sun, and a moon circled her ears. The rest of the planets followed in suit, each rotating in proportion around her radiating form.

She fixed her eyes directly upon mine then, as if to say with great portent, "Pay attention!" With the most graceful, fluid gesture ever articulated, she swept her hand from me towards the window, offering to me everything that was to be found there. Tracing her gesture with my eyes, I found myself gazing out the window onto a great expanse that extended far beyond what my eyes could take in. There were no clouds, there were no planets, there was nothing between me and the great expanse of translucent blue sky that stretched out as far as the eye could see and the mind could imagine. I remained transfixed, staring out the window and engulfed in the peace of that vista, until the woman, the bird, and even my perception of the window vanished.

Several weeks later, while sitting in an introductory art history class, I was stunned to see for the first time, in a work of Botticelli, the face of my visitor in the visage of Simonetta Cattaneo, a member of the Florentine circle and

...i's "Birth of Venus."

...I had a dream. In that dream, I was a
...ner plain and simple bird, something
...e a sparrow. While singing with all my being, I
...pping along a windowsill, poised and ready to fly.

This chapter considers melody and harmony, both relational aspects of music. "Melody is the most readily accessible," says Joanne Crandall (Crandall, 19), as it is made up of a sequence of tones that move in relationship to each other in a kind of dance made up of the elements of pitch, loudness, space, and time. The linear flow of the simplest melody is a living lesson in relationship of the self to the self, and the self to others.

The harmonic intervals, which consist of two pitches sounding together, form characteristic relationships which, in turn, create a particular effect on the listener, ranging (in Western theory) from pleasing consonance to tension-producing dissonance. The illuminating mysteries of the octave have long been extolled mathematically and esoterically, while the interval of the tritone (judged as the most dissonant) was banned by the medieval Church for its displeasing and "evil" qualities. The intervals are seen by some as gateways into deeper consciousness. As mentioned in this chapter, during the Renaissance period, Ficino and others saw themselves as revivers of ancient Greek practices, using the intervals as portals to the hidden realities of universal harmony.

The essential focus of the third chapter is universal harmony. For ancient cultures, universal harmony reached far beyond the theoretical principles and structures of music, while using those same structures as signposts of and avenues to a greater harmonic force at work within creation. Attunement with that deeper harmonic force was considered the primary purpose, as well as the prime generator, of life itself. "The whole of life in all its aspects is one single music; and the real spiritual attainment is to tune one's self to the harmony of this perfect music" (Khan, *Music of Life*, 87).

One of the most striking and defining differences between ancient cultures and "modern society" resides in the way that we understand and perceive music. For the average person living in the twenty-first century, music is, for the most part, perceived as something that comes out of a radio or a stereo system, which can be pleasing or annoying, depending upon one's attitude towards what is being heard. Statistically speaking, very few people actively engage in playing music once they are out of the educational system; music is perceived as having primarily "entertainment value," as something to be enjoyed during time away from work; and, while we say that we like music very much and that it has a profound emotional effect on us, we have lost touch with any understanding of music as being a teacher of life lessons, or as being of divine origin, or as being connected in any fundamental way to the metaphysical cause and effect of our physical reality or bodily existence.

For the Ancients of the Western world, music was grounded in the relationships found in number, within nature. Those relationships of number, in turn, demonstrated deeper, more universal principles found beyond nature. So, it was through music that one was able to understand, yet transcend, physical limitation. The "connection between physical reality and metaphysical principles can be felt in music as nowhere else. Music was therefore justly considered by the Ancients as the key to all the arts and sciences–the link between metaphysics and physics through which the universal laws and their multiple applications could be understood" (Danielou, 1).

In the ancient Vedic society of India, music had always been part of a respect for sound and ritual, in which sound, as the vibratory essence of all reality, was held to be both sacred and the cause of physical existence. In ancient China, music was considered to be a powerful force carefully controlled by the Emperors, as it was believed that music had a strong effect on the citizens of the state.

In ancient Greece, the study of music was regarded as education for the soul; the study of music connected the individual to the World Soul, whose own essence was likened to the beauty and fluidity of music. Grounded in the mythic powers of Orpheus to charm, to enervate, and to overcome even the dark forces of the Underworld through the magic of his music, the Pythagoreans "studied the mathematical principles of harmony that underlie the structure of the musical scale–and which were said to underlie the harmony of the universe as well" (Fideler, "Orpheus," 22).

The strains of Orphic belief in music's power can also be heard throughout Plato's work. In "The Myth of Er," he recounts the ethical considerations of a "good life" in the context of describing a cosmic harmonic progression, which was composed of musical processes by which individual souls entered and exited the realm of the living. In the *Timaeus*, Plato describes the making of the World-Soul, an extremely complex process of numerical and harmonic relationships, which results in the finely tuned harmony of all things and the generation of the visible universe.

In *The Harmony of the Spheres*, Joscelyn Godwin remarks that the *Timaeus* was the only Platonic dialogue never to be lost to the West, even during the Middle Ages, and that this "freak of survival was probably the single most important fact in the transmission of the doctrines of cosmic and psychic harmony" (Godwin, *Harmony*, 4) into Western civilization.

The philosophy of the Music of the Spheres played a

large role throughout the period of the Renaissance. Theoretical speculations concerning music in the celestial realms blossomed. In attempting to revive the Greek musical modes, elaborate systems were developed which connected the sounds of the celestial bodies with our musical scale. Beautiful and elaborate systems of correspondence between earthly elements, the harmonies of the musical scale, astronomical circulations, and spiritual entities in the form of gods and goddesses can be found in the musical treatises from this period.

It is unclear, today, to what extent individuals of the Renaissance undertook a conscious spiritual practice of connecting their own music making to the harmonies of the Music of the Spheres, but we do know that Marsilio Ficino (priest, physician, philosopher, and member of the Florentine Academy) was among those who had developed a type of spiritual discipline in which the tones and intervals of music were employed as a connection to the spiritual forces of the universe. The mystical experiences recounted in the third story occurred to me while studying a stringed instrument whose fundamental technique rests in creating harmonic intervals and they are a reference to Pico and Ficino's work.

It is, however, clear today that we no longer conceive of the universe as a well-tuned musical phenomenon, as we once did, and that:

> the basic concept, as in music, so in life, prevailed in one form or another even up until a hundred years ago. Only during our present century has the belief in music as a force capable of changing individual and society become almost totally forsaken and lost. This means that in the comparative lack of importance which twentieth-century man attaches to music, our civilization stands virtually alone (Tame, 24).

The appearance of the mystical figure at the end of this chapter, summoned through the playing of certain harmonic intervals, may be a reference to several archetypal beings. Harmonia was the Greek goddess who "in one account is said to 'weave the veil of the universe'" (Fideler, 22). The figure also encompasses the muses and Fates who activate the Music of the Spheres in "The Myth of Er," and, in the guise of the sacred dove, she is connected both with the Holy Spirit, the once-female division of the Christian trinity, and the goddess of wisdom, Sophia. Jakob Boehme, a seventeenth-century mystic and music theorist, envisioned the figure of Sophia as the inner essence of "primordial man" (Cirlot, 300), an essence which had been lost, even in his time, and without whom salvation was, and is, impossible. Her appearance reminds us of what we have lost, while being at the same time a kind of guide pointing to the expanse of what is possible through the resurrection of Harmonia.

CHAPTER 4
SONG FLIGHT—Singing Lessons

Shortly after the bird dream I met my singing teacher, a robust man with a bristling energy that surrounded him in everything that he did. He had wonderfully arched bushy eyebrows that stood up and over the frames of his glasses, causing his expression to be one of an eternal question mark. He had moved with his wife and family from the frenetic environment of the big city out into the heart of the countryside, where they bought a barn and some land. They renovated the barn from the ground up, and built inside it a home and a music hall. They lived as simply as possible, off the land and their cows, growing what they needed, making everything they needed, from soap to clothes. They gave wonderful concerts complete with banquets and costumes. He had enough energy for ten people, was always laughing or joking, and he could come up with the most wonderful images that would convey in a flash the essence of the sounds he was attempting to get me to emit.

For instance, if the effect he wanted was for the sound to have a melting quality, he would ask me to imagine a pitcher of hot maple syrup being poured over a pile of pancakes topped with fresh butter. Suddenly, the sound I made felt like hot, sugary syrup, oozing over the layers and making lakes in the crevasses, covering everything with unguent heat and melting the hard edges of the world. If he wanted a sound imbued with lightness and air, he would ask me to picture silk scarves waving in the wind, each one flapping a different color as the breeze played among them, catching them and billowing them out in undulating waves up into the sky. Lo and behold, my dull and heavy sound became lighter and brighter, buoyed up as if it were itself some bright piece of slippery, shimmering silk. If he wanted the quality of energy to move forward in a driving stream of compression, I was to imagine a team of acrobats, tumbling from one side of a stage to the other, head over heels and hand over leg, each like the leotarded wheel of a steam train, one axle connecting to the next and passing the energy on, every completed rotation bringing the group closer to its kinetic goal. As a result, my weakling voice would gather the strength that it needed to get across that stage. Focused and determined, it would leap forward, connecting in precision and power with each of the preordained gymnastic feats required, finishing triumphantly, radiant and strong, with energy to spare.

Along with the images came months of initiatory lessons and endurance training. Most of these involved careful use of the breath and the studied placement of sound within the mouth, with the help of some of the most ridiculous but fun noisemaking imaginable. As compared with the techniques needed for the wind instruments I had played in the past, my voice teacher took breath control to a whole new level.

Breath control meant strengthening the "singing muscle," which was located just below the navel. This muscle could be found by bending over to touch the ankles with the legs straight and placed slightly apart. The main muscle involved in singing is located in the area that dancers call "the center" or Buddhist practitioners called the *hara*. After rising up from the position of grasping the ankles, all other muscles remained inactive, out of the way, and reflexive. The "singing muscle" was the workhorse, and as the singer aligned himself or herself above this central area, the hips were to be long, the small of the back straight, and the chest buoyed up in "poised passivity."

The action of the singing muscle could be compared to that of someone kneading bread in the area just below the navel. To strengthen this area and increase breath capacity, I began by practicing a series of small "puffs," little contracting movements that pushed air out of my body from the region of the center. With my chest and back elevated, the only muscle that moved was the singing muscle as it kneaded the bread, emitting little puffs of air at the rate of seventy-two beats per second on the metronome. This was practiced several times a day, for several minutes each, increasing the number of puffs to be produced.

Then, once certain that the singing muscle was the only muscle truly engaged, I could start making sounds. These began with a "ha-ha-ha-ha-ha," which was supposed to call attention to the looseness of the jaw and tongue. As I calmly surveyed the physical changes within my body, I relaxed my muscles. "Ha-ha-ha-ha-ha." Then came "s-s-s-s-s," a hissing exercise, like the air in a tire leaking out of a tiny puncture hole. Taking in a full breath of air, I made a sound like a leaky tire for as long as was possible–"s-s-s-s-s-s-s-s-s-s-s-s-s-s-s-s-"– in the most efficient way, guarding and conserving air as I squeezed in on that slowly contracting singing muscle. This made me dizzy sometimes. But again, it was practiced several times a day, for a few minutes at a time, and increased my lung capacity.

After being a leaking tire, I then would become a buzzing bee. To find the proper place of resonance within the facial mask, I made buzzing sounds that ranged from the vibrating membrane of a kazoo to the buzzing of a bumblebee. I started with the syllable "oz" to accomplish this feat. Repeating "ozzzozzzozzzozzzozzz," I centered myself in the front part of the mask, resonating an area above the nose and in between the eyes, and I continued to vibrate like a buzzing bee. Then I added an "mmmm-mm" to the "zzzzzzz," gradually turning from a buzzing bee into a cicada high in the tree tops, sounding out for all of nature "zzz-mmm-zzz-mmm-zzz-mmm-zzz-mmm-zzz-mmm."

All of these sounds came before I even could consider forming a vowel! These sounds were meant to relax the throat and they kept my busy mind occupied while I unobtrusively introduced a vowel. To prepare for the vowels, I would sing the syllables "hung-ee." While singing "hung," I lifted the soft palate at the back of the throat, placing the back sides of the tongue in an actively firm position at the back teeth, while allowing the center of the tongue and the throat to completely relax. After sustaining the "hung" for a time, I let go of it to sing the "ee," finding myself automatically in perfect alignment for all the sounds of singing.

Repeating "hunnggg-eeee hunngg-eeee hunngg-eeee," while relaxing and spinning out sound, was a very satisfying and pleasant sensation. It was fun! Air was expelled through the vocal chords in a long steady stream. The voice had only one dimension, length. Spinning itself out in the single dimension of time, it was not wide, and it was not deep. After a certain number of "hung-ees," I would sometimes even find myself in a kind of mild trance–very pleasant and very relaxed, resting simply in the long, gentle threads of softly spun sound.

The lips and tongue then formed the vowels by making different shapes around the air stream, shapes that my voice teacher called "cookie cutters." Those indentations, introduced over the air stream, thus produced the phenomenon of vowels. With the tongue and the lips in the correct shape for each vowel, I got to know each of the sounds intimately. Pure vowels had a clear, ringing quality that neither impure vowels nor the diphthongs had, and finding the essential qualities of the five pure vowels (ah, eh, ee, oh, oo) was a magical thing. Activating an entire series of overtones, each of the vowels opened whole vistas of clarity. Therefore, the aim was to sing as many pure vowels as possible as clearly as I could.

Although there was really only one sound to be made, each of the vowels had its own character or essence. It was necessary to relax into "ee" or it could become tight and complaining. The "oh" was full and round, comforting and expansive. The "ah," on the other hand, was like a long, tall archway, an ancient doorway in a vine-covered wall. As the sides of the mouth came together to create the cookie-cutter shape of "ah," it was as if my mouth was participating in a mystery. The corners of my lips and the flat area of my cheeks created the sides of the door; the resonating mask was the liquid lintel, elongating the archway, and my lower lip and tongue provided the vibrating base of the threshold, over which one stepped as sound passed effortlessly through the vocal chords.

The doorway of "ah" was not a solid construction. It appeared and disappeared, through the vibratory haze, according to the purity of the vowel produced and the clarity of its tone. Similarly, the attendant set of stairs that formed through the mist on the other side would appear longer or shorter, deeper or narrower, depending upon the simplicity of the sound quality created. The doorway hung suspended in the mists of sound, hovering within a constant state of creative flux, responsive both to impulses received from the physical body and to messages coming from the mind, and yet its illuminated pattern remained unquestioningly present and surely attainable within the simple act of truly turning one's attention to its existence.

My favorite vowel was "oo." For me, "oo" was a container of all things. "Oo" had a clarity that produced a brightness and a buoyancy in the body. At the same time, it could encompass darkness, deep and profound. While grounding and ancient, it was uplifting and ever-new. "Oo" could cut through fog. "Oo" had a silver quality to it. At the very top of the arch in the "oo," one could perceive a radiating quality of silver. Waves of moonlight streamed from "oo" as from no other vowel. The waves were long, shimmering strands of silver that emanated in a smooth rope, weaving and twisting themselves–gracefully, gently, smoothly spinning "ooooooooooooo."

I was thinking about the vowels as I left the Past-Times

Used Bookstore one day on the way to my singing lesson. The bookstore was a mish-mash of second-hand books, with topics ranging anywhere from romantic pulp fiction to the mechanics of quantum physics. Loosely arranged by subject matter, the books were all jumbled together. You could find such titles as *In Search of Extraterrestrials* right next to *Vatican II and Its Documents*, or you could discover that *Reincarnation is Making a Comeback*. Pungent odors from the pages of once-owned-but-now-passed-on books clung to the edges of my clothing as I boarded the bus.

I sat on the bus, staring out the window, "puffing" and humming in preparation for my lesson. As the bus rounded the corner of a city block in a ninety-degree arch, I noticed how the sun caught the rounded form of the driver's rear-view mirror. The rays reflected off the top of the circle, radiating yellow-white beams of light, just off the very top of the circumference of the circle. It was like the vowel "oo!"–just like the silver, moonlight beams of "oo," only these were sunlight streams! Holding the images of sunlight and reflecting moonlight within my mind's eye, hoping that the heat of its imprint would clarify my memory, I watched the sound of the sunlight as it streamed off the rear-view mirror of the bus.

My singing teacher was jubilant as usual. We began the lesson with little of the small talk or the traditional exchange of trivialities. The puffs became buzzes, the buzzes became vowels, and the vowels began blending one into another as I relaxed into making the one pure rope of sound.

In a relaxed way, I remained attentive. The singing muscle was working, while everything else was released. The air moved effortlessly through the waves of silken scarves, while I released my mental hold on any negative influences or extraneous thought. In an active stream, the rope of sound moved intensely forward, shaping and caressing the air as it ascended the scale patterns higher and higher. Expressing the energies that lived within each of the musical phrases before me, the musical rope wound its way into the world.

I had been concentrating on remembering those images of sunlight, and happened to be singing the vowel "oo," when suddenly I felt a lifting quality take place within me. It was as if something full of light was pulling up and away from the density of my physical body. Remaining in the delicate balance of that energy exchange, I hovered, suspended yet moving, floating above the physical sounds that seemed to be coming from my own mouth. Poised between the ecstasy of soaring and the practicalities of the sound-making that caused it, I continued to ride the air currents, dipping and diving with each new musical phrase.

I was then aware of the sensation that I had sprouted wings, and that those wings were lifting me up to soar on even higher air currents, generated from some distant place far below. High above the rhythmic intricacies of the constantly pulsating world, I floated freely, the tiny feathers of each wing adjusting to the subtleties of velocity and intensity of the currents. The wings allowed me to hover above the music. I could travel with it; I could travel beside it. I could even enter the current of sound and travel in retrograde against it! Flapping furiously, I could out-race the current. Or, gently gliding, I could rest within its center. Once established in the graceful flow, trusting in the certainty of the sound, the sensation was one of an unending and loving support. The wings that carried me were like hands holding me.

Deep within the limitless variations of this travel, I sensed the quiet and the peace that came from the slow and steady, strong and sustaining beat of those loving wings. I could feel that the wings sustaining me were not really mine. They belonged, at their deepest level, to someone else or some other thing. And from those wings came music. The harmonies that streamed from the breath of those wings fell on ears that perceived them to be colors from the source of life itself. The sounds emitted from those wings issued forth in power from the beginning of time. Caught in the light, transfixed by the sound, I floated on the wind and hung suspended in the air!

I was actually standing on a floor made of cracked, brittle pieces of old red and black asbestos tile, beside a beat-up, out-of-tune grand piano in a musty corner of a church basement, holding a flat and lifeless piece of paper filled with small black symbols placed on five-ribboned staves. Yet, I felt the power of eternal, loving wings generate themselves deeply within the feathers of a single soaring bird. In time the bird, called down by the approaching cadence of the final phrase, folded in its light and feathery appendages and landed with delicate exactitude into the very center of the liquid stream of my final breath.

As I became aware of my surroundings again, I focused my attention on my singing teacher's radiant smile. No words were exchanged, just an understanding. For, as I returned his smile from a deep place of radiance that matched his own, I realized, at least for that day, that my harmony lesson had been complete.

❋ ❋ ❋

In both Western and Eastern musical traditions, singing has long been considered the highest of the musical arts. For the Ancients in the West, the voice was so highly valued because it was *the* God-given instrument, all other instruments being man-made. St. Augustine proclaimed that "He who sings prays twice," indicating the extent to which he felt that song connected the singer directly to the divine realm. Hildegard of Bingen felt that the vibrations initiated in song created sympathetic vibrations within the body of the singer and that this allowed "the words to directly enter the soul" (Gardner, 180).

For those in the East, the "Hindus of ancient times said that singing was the first art [and] the shortest way to attain to spiritual heights" (Khan, *Music of Life*, 93). Through the practice of private chanting, or through public *sankirtans*, singers were reminded of their true nature as sacred expressions of the Creative Word, for it was while singing that the singer was given the greatest sense of joy in a "temporary vibratory awakening [. . . and] in those blissful moments a dim memory of his divine origin" (Yogananda, 158).

This transcendental awareness of sacred sound is spoken of in many traditions. "In many legends the sky hung so close to the earth that it was possible to come and go between the two on an intervening rope (or tree)" (Schneider, *New*, 49–50) reached as a result of singing. In some traditions this is referred to as a silver thread or a golden cord, which, when found, enables the singer to move between worlds or realms of consciousness. The voice, the primal instrument, is the "thread connecting our inner and outer worlds" (Crandall, 101).

The power of the voice itself can be found within the derivation of the word. Our English word "voice" is derived through the Latin *vox* from the ancient Sanskrit word *vaac*, which meant "word" or "speech," both of which were considered to be manifestations of the goddess *Vaak*. *Vaak* is one of the oldest Vedic goddesses and she appears in many books of the *Rig Veda*. Through *Vaak*, "the creative role of the Word seems, therefore, [to be] a notion present from the greatest antiquity" (Padoux, *Vac*, 7).

In the tenth book of the *Rig Veda* (10.125), *Vaak* was worshipped as having created the universe ("In the beginning I bring forth the Father [. . .] it is I that maintain and I that sustain all things in creation"). *Vaak* gave names to everything in existence. *Vaak* was the Truth ("I alone utter the Word of Truth, the Word that brings enjoyment to Gods and men alike"), granting power to those whom She loved

("I make him a divine, a seer, a sage"). She was the inspiration of the seer, both to hear and to see ("In truth, I speak: hear, O holy tradition!"). *Vaak* was consciousness itself, as well as the power of consciousness to be aware of itself; she was inherent in communication of any kind, but especially in sacred ritual chanting.

The power of the vowels is also spoken of in this story. Worthy of being worshipped, and capable of transmitting power, the resonant qualities of the vowels have, for centuries, been perceived worldwide as portals or gateways into sacred reality. "Graeco-Egyptian papyri speak of the ruler of the gods, the king Adonai, as 'Lord I-A-O-U-E-Ae,' while some of the oldest *mantras* of Egyptian origin use[d] only vowels" in the context of initiated mystic circles (Hamel, 115). The ancient Hebrews considered the vowels as connected directly with the holy wind, *ruach*, and thus far too holy to be represented by written symbols; for centuries there were no vowel pointings in the Hebrew texts, as the vowels were considered too powerful and too sacred to be consigned to an image.

Religious rites worldwide have always been accompanied by chanting, and song itself has been used specifically for spiritual restoration and healing all over the globe. There are countless *mantras* found in the Vedic texts of India, which are to be used for specific healing purposes. In the Old Testament, we are told that David was summoned to sing and play his harp when King Saul was taken ill. Song is used universally by Native Americans for healing purposes.

There are several stories in this book that allude to connections between birds, spirituality, and music, the first of which is found in Chapter 4, where the transcendental sensation of being transported into an alternate reality while singing is described. Bird imagery, connected with the phenomenon of sound, is found in the folklore and customs of many cultures that honor sound or music as a transcendent medium between this ordinary world and the magical "other worlds" of divinity.

In ancient China, the story goes that God sent music down to earth by sending a pair of whistling birds to Ling-Lun, who then translated the melodies received from those divine emissaries into our earthly music (Hamel, 87). In the Mayan culture, *Itzam-Yeh*, "the Bringer of Magic," was a cosmic bird who symbolized the forces of life and death, union with whom was sought through shamanic states of rhythmically induced trance (Gillette, 10). In ancient Egypt, each person was considered to have seven souls, one of which was called the *ba*; the *ba* had a bird's body with a human head.

The folklore of medieval Celtic Christianity is filled with spirituality linked to the musical imagery of birds. There are many tales about the travels of saints to musical islands or musical lands under the sea. "The Celtic legends leave an unforgettable impression of a dreamlike existence suffused with music: an Elysium of musical trees, fruits, fountains, stones, nets, choirs, and birds upon birds" (Godwin, *Harmonies*, 55).

In *Sound and Sentiment*, Stephen Feld describes the social customs of a traditional society that is still very much in touch with its ancient customs, the Kaluli people of Papua, New Guinea. As a rainforest people, the Kaluli use sound over the other senses to negotiate their everyday lives, where they are surrounded by a world of birds and birdsong. In an intricate and beautiful cosmology, the Kaluli people believe that the birdsongs that come to them from the forest are the songs of the souls of their dead ancestors. "To you they are birds, to [us] they are voices in the forest" (Feld, 45). During elaborate funeral rituals, during which the women weep and call like birds, spirit mediums don feathers to "become a bird." To the Kaluli, "becoming a bird" means to pass from this life into the next. Through the use of sound and rhythm, they enter a state of trance, from which they can communicate with the spirit of the person who has just died, sing his or her songs, and tell the mourners where to look for him as a bird and in which trees. "Bird sounds metamorphose Kaluli feelings and sentiments because of their intimate connection with the transition from visible to invisible in death, and invisible back to visible in spirit reflection" (ibid., 84).

The sound traditions of India are also rich with bird imagery. In the *Rig Veda*, the sun, *Surya*, is often referred to as a "fine-winged (*suparna*) bird." The Vedic fire altars were often built in the shape of a bird, with its wings spread in flight, while the meters of the verses chanted into the fire were also referred to as birds, being "speedier than horses" (Dange, 43) in connecting the chanter to the heavenly realms. The *Gayatri* meter is specifically mentioned as being like a bird.

In later scriptures of India, the sacred syllable "om" or "aum" is often described as a bird; its right wing consists of A, its left wing consists of U, its tail is comprised of M. It is said that "an adept in yoga who bestrides the *Hamsa* (bird) thus [. . .] is not affected by karmic influences" (Staal, 126). The *Hamsa* (a goose) is also found in the well-known great *mantra* "aham":

> If the *mantra* "aham" means "I am THAT," it may be put into Sanskrit as "so-ham," and this, in fact is frequently done. The mirror image of "so-ham" is the *mantra* HAMSAH. Since the word *hamsa* refers to the mythological gander long taken to symbolize the Self, this *mantra* may be translated "[I am] the Cosmic Bird" (Alper, 281).

The *Rig Veda* speaks of two birds, "friends joined together," thus illustrating "a key Vedic idea: that the individual spirit and the universal spirit both are of the same essence" (Feuerstein, 132). This scene is paraphrased in the *Mundaka Upanishad*:

*Like two birds of beautiful golden plumage,
inseparable companions,
the individual self and the immortal Atman
are perched on the branches of the self-same tree.
The former tastes the sweet and bitter fruits of the tree.
The latter remains motionless, calmly watching.*

*The individual self, deluded by forgetfulness
of its identity with the Divine Atman,
grieves, bewildered by its own helplessness.
When it recognizes the Lord, who alone
is worthy of our worship as its own Atman,
and beholds its own glory,
it becomes free from all grief.*

CHAPTER 5
THE WORD MADE FLESH

For several years after I graduated from college, my life seemed to tick along at a fairly regular rate. I was able to make a small living at music, supplemented by some work as a typist. I was involved in a relationship that seemed to be headed for the altar and life's little routines simply followed one after another like a series of pre-planned events set to unfold on automatic pilot.

Gradually, however, the certainty that I had come to expect began to dissolve. At first, it seemed as if the timing of things was just off, like so many bad connections or missed cues. Eventually, I watched, as if perched in some scientifically detached observation booth, as one thing after another on my docket of best-laid plans fell apart. The way, which once had been so clear, became clouded. Events that had been certain were vanishing into thin air. The intimate relationship that had once been so sure had faded, and I watched as each of the foundational pillars that had supported my world disintegrated, crumbling into dust beneath me.

For the first time in my life, I felt the full weight of being completely and unquestionably alone. The sensation that there was absolutely nothing to hold on to, no one to turn to, engulfed me. Adrift, with no direction in a sea of circumstance, I felt that I would lose myself forever in the great, dark morass of existence. From somewhere deep inside myself, I screamed out, begging for someone or something to catch me, before I fell irretrievably into the great black hole whose mesmerizing inertia was beginning to swirl menacingly around me. My inner voice rang out in sheer despair, its sound bubbling up from underneath regions of inner pain. It reached out to whatever in the universe might respond, stifled only by the hardened will power of my physical body, which would not let it escape—yet it silently resounded to the ends of the earth and shook every fiber of my being.

They say that when you come face to face with something is when you finally see. They say that when a student is ready, the teacher will appear. They say that when you truly turn to God, then God will truly hear. This is a story of just how ready I was, and how by turning to listen, I found that I had been heard all along.

For months, I had been using the kitchen table as a desk for writing papers or copying music. I had been vaguely aware of a small, unobtrusive sound that the stove would make from time to time, but it was never loud enough to demand my full attention, so I simply allowed it to remain in the far reaches of my consciousness. Maybe it was that my fiancé and I were breaking up after three years that kept me from focusing on anything other than how I was feeling, which was not great. Three years may not seem like a long time to some people, but to me it seemed like an eternity, and the breakup was an unimaginable ending to a whole world of shared friends, jointly owned possessions, and common living space.

One of the final acts of kindness and concern that my fiancé showed to me as our relationship inched to a close was to call the power company to come and take a look at the stove in the apartment, since during one of his final visits to divide possessions and to re-examine "what went wrong" for the hundredth time he thought that he had smelled gas in the apartment. I had been oblivious to any such smell, but agreed to let Consolidated Power and Electric come and investigate the next morning, which would have been a Monday.

That Sunday night, after he had left and I was again engulfed in the pain of being alone, I tried as best I could to describe to my mother on the telephone just how desperate I was feeling. I remember telling her that I thought I might start reading the Bible, that maybe there would be some comfort or some answers to be found there. This must have come as somewhat of a shock to my mother, but she did manage, as a woman not much taken with most religion, to choke out a supportive response: "Well, dear, they do say that it *is* a good book." So, I scanned the myriad titles on my bookshelves for that dusty old copy of the Bible that I had transported from one place to the next since childhood but had never opened. Leaving it next to the bookcase, I resolved right then that sometime soon I would begin reading it in earnest.

The next day, on Monday morning, I let the man from Consolidated Power and Electric in the front door and up the stairs to my second-floor apartment. "My boyfriend–well, I guess I should say my *ex*-boyfriend–thought he had smelled gas in the apartment," I explained, and went back to working on the papers at the kitchen table. The man from the power company poked around at the stove for a while. He examined each of the burners, opened the oven door, and reached down behind the oven racks.

"Oh yeah," I added. "I have noticed that the stove has been making a funny noise of some kind, but, you know, just every once in awhile."

He continued working on the stove, inspecting every connector and piece of metal he could find. Finally, he stood straight up, hands on his hips, and said, "Lady, you are one lucky individual! You could have been blown to kingdom come at any time! This stove has been turning *itself* on and off. There's an opening in one of the lines, and it's been coming on and going off all by itself. On behalf of the city, I declare this stove *condemned!*"

"Condemned?" I asked. "But who takes care of getting another one?"

"Well, I guess your landlord will have to provide one, but we can't leave this thing here. This whole place could have gone up in smoke at the drop of a hat! I'll be right back." He bent down to turn off the main gas spigot, then walked out through the dining room into the living room, and disappeared down the front stairs.

He returned shortly with a bright red tag that declared "CONDEMNED," and after signing his name and the date on it, he attached it to one of the openings at the back of the stovetop. Then, taking out a large roll of orange duct tape, he unrolled two long strips of tape and formed a huge "X" across the door of the oven, indicating that this stove was not to be used. He finished by disconnecting the stove from the gas line.

By now, my head was spinning. I was beginning to realize that that "small, unobtrusive" sound I had been hearing for months was the sound of the oven coming on all by itself. I was beginning to see that I had been sitting at the kitchen table for all those nights, right next to the stove, unaware of gas leaking, unaware of the pilot light, and that at any point I could have been killed instantly in a sudden explosion. I was realizing that the only difference between life and death was but a second's worth of happenstance and some good luck. I was realizing that this was the last straw in a long series of unspeakable events. That I was even standing in that kitchen alive was itself a miracle!

My mind and my body were numb. It was as if they weren't even really attached to each other anymore. I was nonplussed as my mind spun around in a numbed haze. My hearing, on the other hand, seemed to have become extremely acute. I was becoming aware of some very unusual sounds. I realized that the man from the power company had been humming some very beautiful melodies the whole time that he'd been in the apartment. He had been humming on his way into the apartment; he had been humming as he worked on the stove; and he was humming now, in a beautiful, warm voice, as he continued to finish the paperwork that he needed to fill out for the stove.

"What are you singing?" I asked as I watched him write.

"Oh, I was on retreat this weekend with my church. It was Pentecost yesterday, you know, and we spent the whole weekend singing and praising the Lord."

"Oh," I said, focusing on his voice.

"Do you know Jesus?" he asked, raising his pen off the page.

"Well . . . I've heard . . . a lot about Him," I responded slowly, "but I can't say that I really *know* Him," as I thought to myself that, after all, He *was* dead.

"Oh," the man from the power company said. "Well, do you have a Bible?" he asked, indicating that he would like to see it.

"Yeah," I answered. "There's one upstairs. I can get it," I said, and went up the attic stairs to find the Bible in the spot where I had left it unopened from the night before.

When I handed him the book, he opened it to Romans 10:8, and he pointed with his finger as we hunched over the book together. We began reading, "The Word is near you, on your lips and in your heart," and as we read together, I slowly began to become aware of a tangible presence in the room. Something or someone was filling the space in my living room with a presence that was alive, and from there it was radiating its way into the empty space in my heart, with a deep warmth and a sense of caring that I could feel in every part of my being. The "Word" had taken on breath, was coming to life, and the printed letters on the page were joining with the forces of the world and they were both right there in the middle of my living room and their energies were enveloping me and caring about me.

As I listened to the printed words "the Spirit helps us in our weakness," there came the sound of a great wind that filled my ears as it swept through the room, like the sound of a tornado or the passing of a freight train, and within the swirling of the wind, the words I heard were so much more than ordinary words. The words I was hearing were covered with flames. Each of the letters seemed to be made of feathers, feathers made of flame. The words were made of fire, and the fire was generating energy into the wings, and the feathered wings sent them through the air. They came, traveling by the medium of the air, but it was the fire that sent them. "And he who searches the hearts of all knows what is the mind of the Spirit, because the Spirit intercedes for the saints according to the will of God."

I stared at the man reading the Bible to me in my living room. His words were on fire and time was standing still. Time had come to a rushing halt in that very instant. I saw my entire life before me, like so many images shaped and held afloat in a movie projector's stream, unreal, and yet more than real. I saw every event that had brought me to this point, but I was seeing my life's movie from the point of view of the director. I saw how it all had been planned from one scene to the next. This was the way it was supposed to happen. It had all been orchestrated, and each and every life event that had never made sense before, that had been so painful, that had literally almost blown up in my face, was all meant to bring me to this very moment! And, while realizing this, I also knew that I had been waiting for this point of understanding my whole entire life! God and the forces of the universe may have had to conspire to blow me off the face of the earth to get my attention, but God certainly had my attention now!

From this new vantage point, I could see a lot of things. I could see how fragile life was. I could see how my very existence was dependent upon forces beyond my control. I could see how, with God, anything was possible. I could see how it was possible that Jesus had been sent to die, even for me, centuries before I could possibly know His name! I could feel the tremendous love with which He had been sent. I could see how simply believing that He had been sent by God with such deep love carried the power to set me free!

I could feel the possibilities, because I could feel the love. It was tangible in the room. I could feel, all around me, the caring of God and the presence of Jesus, in that moment of time, which had come to a screaming halt in my living room. The holy wind that had surrounded me began to subside. The heightened sounds that I had been hearing began to wane and to reclaim their normal proportions as ordinary words as I internally answered "yes" to the possibility of God's loving even me.

I answered "yes" to that moment. I answered "yes" to the prayer that God's angels would surround and protect me. I answered "yes" to the new beginnings of God's invitation, and the man from the power company left my apartment, walked down the front stairs, and, presumably, continued on to his next customer.

For weeks, I wandered in a state, blinded, I suppose, like Paul when called into the city, speaking with those who could lead me, and worshipping with those who could inspire me. I combed the Scriptures daily for guidance and direction and I listened for the way to go. My "messenger" from the power company visited me a couple more times, and each time he did, the pace of normal events would mysteriously slow down, my awareness would sharpen, and my hearing would become piercingly acute, but soon his message was delivered, and we lost touch with each other over the weeks.

Four months later, on a crisp fall day, late in the afternoon, I stood beside a small country pond, dressed in a long white gown and surrounded by a few close friends. I took the hand of the woman who was to become my mentor in the ministry and we walked out into the cold, still water of the pond. I looked down to see the many small fish crisscrossing and darting around my ankles, and I thought of all of those who had gone before me, and all of those who would follow. I felt surrounded and centered in a great stream that was centuries old.

As my body was lowered down, the cool, autumnal waters rushed in around me, engulfing me in waves of clear, sparkling liquid. The waters poured over every inch of my body, separating things that had been from things that were to come. They startled my physical senses and brought my mind to complete attention. Alert and present only in that one glorious moment, I felt the power of love and forgiveness cleansing the deepest recesses of my world-weary soul.

The waters of purification, ancient and divine, washed over me and held me in their arms. Finally, I heard the words that came in answer to my desperate, internal prayer of months before: "In the name of the Creator, in the name of the Sustainer, and in the Name of the Redeemer, you are my beloved daughter, with whom I am well-pleased."

Somehow, somewhere, beyond the events of everyday life, my frightened and paralyzed prayer had been heard. And, more miraculously, within and through those self-same everyday events, my prayer had been answered! Disaster had become salvation. Straining to hear, I had been heard.

It was true, what they say. It really was a whole new life, and I was singing a whole new song!

☀ ☀ ☀

In this chapter, I share the story of an encounter with the living Word. It is an account of how the sounds of that Word can have an active and physical effect if one is willing to listen when "the Word draws near." While the story, as any story, is culture-bound in language and time, the creative Word is as old as civilization itself. The Word has been sounding, in many ways and in many forms, from the beginning of time.

The Word, as such, is both Creator and creation; it is

vibration, especially represented in the phenomenon of sound. "According to all spiritual sciences, first God created sound, and from these sound frequencies came the phenomenal world" (Johari, 36). The Word, thus, has both unmanifested and manifested aspects; it begins as the unmanifested Being of Deity, which then becomes, in stages, the manifested world of creation.

Ancient cultures from all over the globe have universally expressed belief in the powers of the creative Word. India's Vedic culture was one of the first to place a heavy emphasis on the power of sound.

> In the Hindu theory of vibration, matter is the most "condensed" of vibrations; it is solid and perceptible to the senses. Energy is less rigid, more subtle; [. . .] it is still patterns of vibration, only in a more subtle state. The subtlest of vibrations, according to the ancient sages, is the so-called cosmic sound, the creative Word out of which the entire universe of stars and seas, plants and animals and human beings has evolved (Easwaran, 53).

In other words, God is Sound; God is existence; existence is Sound.

We focused above on one of the oldest personifications of the creative Word, the *Rig Vedic* goddess *Vaak*. In some passages, *Vaak* is Creator; in some She is Co-Creator. In all, She is the Universal Sound.

"Early Hebrew mythology had a copy of *Vaak* in the female deity *Bath Kol*, 'Daughter of the Voice,' who was the source of the prophets' inspiration" (Walker, 329). Likewise, in Hebrew scripture, the *dabar*, or Word of the Lord, figured very prominently. It was through the sound of the Word that the Lord spoke the universe into being: "Let there be light" (Genesis 1:3). It was through *dabar Yahweh* that Abraham was called to leave his homeland, with the Word of the Lord coming to him in a later vision (Genesis 15:1) to make a covenant with him. It was the powerful Word of the Lord that spoke to Moses on Mount Sinai, giving him the Decalogue, or the Ten Commandments: "I am the Lord, your God, who brought you out of the land of Egypt, out of the house of bondage" (Exodus 20:1).

The Word of the Lord came to the prophets of Israel: "Hear, O heavens, and give ear O earth: for the Lord has spoken" (Isaiah 1:2). It is through the prophets that the Lord's message was to be delivered, as they were the foremost receivers and transmitters of the Word: "And I heard the voice of the Lord saying 'Whom shall I send, and who will go for us?" (Isaiah 6:8-10).

Dabar Yahweh, not surprisingly, was often connected with the breath of God, as it was the breath of God that supported the words of creation once they had been uttered. In the *Zohar* of the thirteenth century (one of the greatest of the *Kabbalah* texts), the main character, Rabbi Shim'on, taught that King Solomon had gained access to the mysteries of the universe by learning the divine breathing techniques from his father, King David (Abram, 247).

In *The Spell of the Sensuous*, David Abram makes the point that the four letters that make up the most holy of God's names–YHWH–are, in fact, those consonants that are the most breath-like, thus, indicating that that which is most sacred is really only capable of being spoken by the wind (ibid., 249).

> Some contemporary students of *Kabbalah* suggest that the forgotten pronunciation of the name may have entailed forming the first syllable, "Y-H," on the whispered inbreath, and the second syllable, 'W-H,' on the whispered outbreath–the whole name thus forming a single cycle of the breath (*ibid.*)

As found in the Hebrew tradition, *dabar* offers a rich and

profound understanding of the Lord's relationship to creation, ranging from the very essentials of creation, to being a guide for living within that creation (the Law given to Moses), to providing inspiration to the prophets, to being the Scriptures themselves.

The ancient Greek world also taught a highly developed philosophy of the Word in the doctrines of the *Logos*, dating back to Heraclitus in the fifth century B.C.E. and the Stoics of the fourth century B.C.E. (Achtemeier, 498). The original term *logos*, which has come into English as "word," and which has typically been taken to mean "reason," had many meanings in its original usage. It could have held the following meanings:

> 1) order or pattern; 2) ratio or proportion; 3) oratio, a discourse, articulation or account, even a "sermon"; 4) reason, both in the sense of rationality and in the sense of an articulation of the cause of something; 5) principle or cause; 6) a principle of mediation and harmony between extremes (Fideler, *Jesus*, 38).

Understood in so broad a context, the *logos* of the Greeks was indeed similar to other cultural understandings of vibration, for instance that of *vaac* or *dabar*. Through a kind of divine discourse, a reasoned and well-proportioned harmony was seen as underlying everything. "Logos, as a principle [was] the natural order of things, the principle of reason, relations, and harmony, [. . .] both within the natural fabric of the universe, and within the human mind" (*ibid.*). A primary force in Greek cosmology, the *logos* was a mystical, divine force that "encompassed all of these meanings and refer[red] to the underlying Order of the Universe, the blueprint on which all creation is based" (*ibid.*).

Unlike the Vedic culture, which concentrated on grammar as the most important aspect of the Primal Sound, the Greeks focused on mathematics and the numbers of that

Sound. In the West, it was number that led to all mysteries, as well as to a sense of justice: "That's why Plato insisted that the guardians of his ideal state, described in *The Republic*, needed to study harmonics, for a mind exposed to the principles of proportion is most likely to have a true appreciation for the nature of justice, the principle through which 'every part receives its proper due'" (*ibid.*, 22).

From this, it is easy enough to recognize strains of these earlier influences in the Christian theology of the Word. Christianity is a religion of admittedly syncretistic tendencies, and roots of both Hebrew theology and Greek philosophy are deeply embedded within the books of the New Testament. Portrayed as a prophet, Jesus was a preacher of the Word in the synoptic gospels (Matthew, Mark, and Luke); outside of the gospels, the "good news of Jesus Christ" was referred to as the Word of God; and finally, in the book of John, Jesus himself was given the salvific title and actually proclaimed as being "the Word of God." The concept of the Word is central to the book of John, indicating both "possibly early Gnostic influences" from the Greek world, as well as a "profoundly Jewish substratum" that resonates with the scrolls found at Qumran (Achtemeier, 498).

There is a profoundly beautiful theology of both Sound and Silence woven throughout the whole of the New Testament. It is grounded in the Hebrew understanding of *dabar*, but develops the concept that, in speaking the Word, God makes the choice to become silent. It is the lowly Mary who is able to hear God, as she answers "yes" to the possibility of bearing the Word within herself. "Mary's poverty, which allows her to be open to God, is characterized by silence" (Gawronski, 113) as she accepts co-creation with the Lord.

Jesus is born, the Word of God, and during his life he speaks God's truth to those "who have ears to hear." As the embodiment of God, now silent, Jesus is the Word-seed sown in the world. He is the Word-seed that interacts with the "field" of manifested existence, affecting and changing that field through the very fact of his existence (Gawronski, 119). However, the Word becomes mysteriously silent upon the cross. God is also inexplicably silent to His Son during His time on the cross.

This is the "crux" of the matter, for it is precisely at the point of suffering and death that Jesus must remain silent and alone, because "Jesus' last cry is solidarity with all of suffering humanity" (Gawronski, 100). In the Word's last utterance, He identifies completely with the human situation, entering into the great silence of death, as does every other human being. So, for those who have ears to hear, "the Cross is not the world's last word, but it is God's last word about Himself" (Gawronski, 105). In the silence of the cross, a new relationship was forged. God was willing to come into the world as the Word, out of love for humanity. It is in the resurrection that we understand the silence of the cross not to be the last word; God's love and new life have the final say.

A theology of the Word is also found within Islam, within the "the doctrine of *ism e-Azam*, which can be interpreted as the doctrine of the mystical word" (Khan, *Music of Life*, 228) as well as being witnessed in the deep reverence given to the words of the Koran. Respect for the Word is found within ancient Zoroastrianism, as well as being found in the theologies of many indigenous peoples. The spiritual sound path referred to as The Path of the Masters also places great importance on the Word. Using the Sanskrit term *shabda* (word, sound, voice, speech), the many generations of spiritual teachers speak inspiringly of the Word as being a sound current connected with the Holy Spirit, sustaining "millions of universes and regions. [. . .] It is the soul-current of consciousness. It is the Celestial Melody, [. . .] *Shabda* is a string which connects everyone and everything with the Lord" (Singh, S., IV, 120–21).

An overview of indigenous religions and ancient theologies results in a stunning and universal cosmology that reflects a reverence for the vibrations of creation articulated as the Word. Likewise, the universal perception of the Word as life-giver and sustainer is perhaps best illustrated by the appellation given to the sacred syllable ("om") itself. Called the *pranava*, the "ever-new," it holds within its very name the spiritual attributes which it is believed to hold and to give–creativity, refreshment, and newness of life.

CHAPTER 6
HOSPICE PRAYER

The Spirit who moved me to baptism also moved me to put my faith into action. The sonic force that had called me into the ministry also pulled me into hospice service as chaplain to the dying. While I had attended seminary, and while I had been through a lengthy process to become ordained, little in my formal training could have prepared me for the intimate and deeply moving relationships that would develop over the years with the hospice families whom I served.

My first chaplaincy, right out of seminary, proved to be baptism of another kind. It was baptism into myriad geographic areas–the city, the farm, the suburbs, the slums. It was baptism into the many ways that people live their lives, baptism into the family dynamics of how people can hurt one another and yet heal one another, baptism into how people can love one another. It was baptism into the fires of mental suffering and physical pain. It was baptism into the heart of the world.

Many of the people I met were older, and most of them had lived very full lives. While parting from their families was certainly difficult, there was at least the satisfaction of knowing that a life had come full circle, that a pattern had been completed. It was not so natural, however, when serving those who were younger, those who, in the prime of their lives, found they were succumbing to irreversible disease and would have to say goodbye to children who were still growing and who still needed their parent's love. Those were the situations that were most difficult. They demanded all the energy that I had and called on every resource I could summon.

I met Joan when she was thirty-six. She had been diagnosed with breast cancer several years before, the treatment for which had not stopped the cancer from spreading to her lungs, to her bones, and to her brain. She had three children, ages fifteen, twelve, and three. Her ex-husband, an alcoholic who had been abusive to her, had left the family during her last pregnancy. He was no longer in the picture.

Joan had been a very pretty woman. You could see it in the pictures displayed around the house. You could see it in the faces of her children. But now, following the ravages of the disease and several courses of chemotherapy, her face showed the strains of unnatural aging brought on by trauma. Once she had had striking red hair, but now her head was covered with a light blonde peach fuzz, as her hair attempted to return after chemotherapy under the cloth turbans she wore to camouflage her many losses. She had also gained a large amount of weight from the drugs meant to reduce the swelling in her brain.

With all of the treatments and with her many physical and emotional losses, Joan had been through the mill. Or, put more truthfully, she had been through hell. When she was referred to our in-home hospice program, Joan was still in the hospital receiving chemotherapy treatments. The treatments, however, were not helping. She was struggling with whether to discontinue treatment and relieve herself of the constant nausea, or to continue ingesting the poisons that made her as sick as a dog in order to buy possibly a little more time with her children.

The day I made my first visit, she was weak from retching. She had sent the children home, since watching helplessly while she was continually sick upset them, and concern for them was an additional drain on her. Tears of anger filled her eyes as she explained how tired she was–she was sick and tired of being sick and tired! Every few minutes as we talked, she would prop her weak, almost limp body up onto one side with her elbow, reach for the emesis basin, and attempt to vomit. But, with nothing left in her system, her body was simply wracked by waves of dry-heaving, her skin glistening with sweat, her throat rough from gagging, and her eyes wet with prolonged pain. Then she would fall back into the bed, exhausted, inhaling in short, labored gasps of air, unable to move among the rumpled bedclothes, tendrils of medical monitoring lines, and the small piles of used, crumpled tissues.

I helped her take small sips of water or little bits of ice, enough, hopefully, to ease her parched throat and mouth, but not so much as to bring on another bout of nausea. I took the warm washcloth that had fallen to the side of her bed, rinsed it out, filled it with cold, clear water from the bedside sink, and placed it again on her heated forehead.

"What did I do to deserve this?" she cried out, rolling over to challenge me directly. My stomach knotted into pain and my own eyes filled with tears as I sensed the fierceness with which her mind, in its search for an answer, relentlessly blamed herself for having done something wrong, or for having done something to disappoint God so that this plague had been sent down as retribution upon her.

"Joan," I said, taking her hand and wondering what could be said that could bring any relief. I gazed back into her eyes. "You didn't *do* anything. This is *not* your fault."

"I am so tired. I want to stay with my children; I want to see them grow up. They need me. But I am so tired of fighting–I just can't do this anymore. I have got to have some peace!"

The hospital room had a cold and impersonal feel; it was filled with noisy machines and medicinal smells. But her utterance of the word "peace" took me back to an old seminary classroom, to a place in my memory where we had studied our prayer book and the liturgy for "Visitation of the Sick." Centuries ago, when a priest visited someone who was sick, he entered the house with the words "Peace be unto this house, and to all who dwell herein."

We still use similar words in our prayer book today: "Peace be to this place, and to all who dwell in it." We no longer focus on sickness as retribution from God, as they did back then; today we focus on God's presence and forgiveness. The inclusion of that one sentence, "Peace be to this place," speaks to the abiding human need for God's peace, especially in times of sickness.

We had even learned some of the Gregorian melodies from that early liturgy (part of the medieval Sarum rite), its use having stretched from Canterbury throughout England, even to Ireland and Wales, before the blending of Catholic and Protestant elements, which wrought so much change. It was unfortunate that the Reformation had resulted in the loss of so much music, as well as the sacred custom of singing many services. There were beautiful, calming melodies from the services for the sick that the priest could have sung. Used for centuries, they had been consigned to oblivion during the periods of reformation. There was the simple melody, "*A porta inferi,*" which speaks lovingly of God freeing us from the darkness. There was "*O Oriens,*" an antiphon calling for the coming of divine light. There was the lilting "*Asperges me,*" taken from the Psalms, that tells how God can wash us clean, completely clean, in a purification beyond our own depths. There were so many beautiful and soothing melodies.

As my attention came back to the hospital room, I realized I had been repeating a melody for some time. I had been singing it, over and over again, like a *mantra*, repeating it in an even, measured manner. "Peace be to this place, and to all who dwell within it," I sang. I looked down to see that I was still holding on to Joan's hand. "Peace be to this place, and to all who dwell within it," I repeated, spinning out the melody slowly and soothingly, like a web of sonic cotton that wrapped all around Joan's body as she lay in the bed. By now she had released herself back onto her pillow. Her eyes were closed and she was simply listening calmly to the repetition of that simple melody. "Peace be to this place, and to all who dwell within it." The melody had seemed to bring a sense of calm into the room. For now, at least, the nausea had subsided.

"Joan, I'd like to form a breath prayer with you," I said. She nodded her consent. "Simply imagine that you are standing in an open and peaceful field, that you are standing before the loving presence of God in that field, and that God is looking down upon you and smiling. Now, if you were to address God, what would be your favorite name for God, your favorite way of addressing God?"

Joan took a few seconds to reflect on what that might be and then answered, "Our Father in Heaven."

"So," I said, "in a very short prayer that doesn't use a lot of syllables, you might address God as 'Heavenly Father'?" She answered "um-hm," nodding "yes" with her eyes still closed.

"Now, if you could look directly into God's eyes and ask for the thing that you desired most of all, in your heart of hearts, what would you say to God, and what would you ask for?"

"I would tell Him how tired I am, and that I just want to find some peace," she replied quickly, shaking her head back and forth, as if in disbelief.

"So, perhaps this breath prayer that we're making for you would go something like this: 'Heavenly Father, grant me your peace'?" I asked.

"Yeah, it would go like that. Just like that."

"Okay, Joan, when you are feeling most tired, and when you're feeling really alone, then I'd like you to try repeating this breath prayer. Take a deep breath in, and as you let out your breath, simply repeat the words of that prayer. 'Heavenly Father, grant me your peace.' You can say the prayer out loud, or you can say it silently to yourself. 'Heavenly Father, grant me your peace.' Repeat it over and over again, in rhythm with your breathing. 'Heavenly Father, grant me your peace.'"

I told her that this was her very own prayer, that she

had fashioned it herself, and while she could have other people repeat it with her, it would always be her very own prayer, which God would know and hear.

"Will you try using that prayer for awhile?"

"Yes," she answered. "Heavenly Father, grant me your peace. Heavenly Father, grant me your peace. Heavenly Father, grant me your peace."

I explained as I wrote down the prayer on a piece of paper that I was titling it "Joan's Prayer." I placed the piece of paper with "Joan's Prayer" on her bedside table, and told her that she could use the prayer anytime, and that if she wanted to, she could share it with members of her family or with the staff, or maybe her mother would even pray it with her.

I asked whether there was anything else I could do for her that day. "No," she answered. "Actually, I'm feeling much better now, thank you." And, at that point, we both saw that since we had started the meditational chanting and the breath prayer she had become much calmer. A kind of peace had settled over her, and the vomiting had stopped.

Over the next few weeks, I visited Joan in her home usually once a week and we continued to use her breath prayer and some gentle chanting of melodies during our time together. Sometimes all the children, as well as Joan's mother, would join us, and seated in a circle with Joan as the central focus, we would say a prayer of thanksgiving for God's presence, ending with everyone chanting together Joan's special *mantra* for God's peace.

By now, Joan had decided to discontinue treatment, and while she was free from nausea, she had been having some trouble breathing for awhile, so eventually, again, she had returned to the hospital. She had signed the documents against using extraordinary measures to keep her alive should one or more of her physical systems shut down. She had had enough of being poked and prodded, stuck and stung, and she was committed to allowing things to unfold naturally, trusting in her oncologist's ability to keep her free from pain.

I was called early one morning by one of the hospital nurses, who reported that Joan had taken a turn for the worse, and that she wouldn't live for more than a couple of hours. Joan had asked that I be called to come be with her and the family.

When I arrived at the hospital, everyone was sitting or standing at different places around Joan's hospital room. They were obviously upset and had been or were crying. I asked everyone to gather around Joan's bed and join hands. We prayed a prayer of thanksgiving, for everything that Joan had given to each of us and for everything that she meant to each of us. Then we prayed that God's loving presence, in the form of His living light, would fill Joan from the top of her head all the way down to the bottom of her toes. We prayed that she could feel God's love filling every part of her; then we closed with her own prayer for peace.

Joan was barely conscious at that point, but she did open her eyes to gaze at me and smile. I asked her if she could feel the Light, and she nodded "yes." I asked her if she could feel God's peace, and she nodded "yes."

I quietly asked the children if they wanted to say anything special to their mom at that time. Slowly, Carrie, the fifteen year old, thanked her for teaching her all the things she had taught her. Joan's mother picked up Jessica, the three year old, and held her close to her mother's cheek. Jessica reached over and kissed her mom, saying brightly, "I love you, Mommy. I hope you feel better soon." Then Joan's mother, having given Jessica to Carrie, also said her final goodbyes to her daughter, giving her permission to leave this world of pain and to rest in the peace of God's loving arms. Only Jeff, the thirteen year old, quiet and sullen, seemed unable to muster the words or the feelings to speak.

We all stood or sat on the bed, hugging each other or holding hands. Joan's breathing had decreased to about four respirations per minute. She was just barely in this world, and you could sense that her spirit had largely left her body. With her goodbyes said and with permission to go having been given, Joan was in the process of letting go. With tears in our eyes and wrapped in stunned silence, we remained suspended, holding on to each other and holding on to Joan, attempting to help her stay peaceful and calm, as we watched and listened to the ever so faint puffs of air that came less and less frequently, until, finally, there was nothing . . . just . . . silence.

We were all floating in that mesmerizing sea of air as it was gently calmed and quieted, having been enveloped into the higher seas of the ethers of peace, when suddenly, Jeff let out a tearful, blood-curdling scream–"No, Mom! Don't go!!"–and he threw himself on her body, wrapping himself around her neck and shoulders.

Startled, my entire body clenched in shock. Everyone in the family began wailing and crying again, and Joan, who had actually left her body to make the transition, inhaled a huge, raw gulp of air, with every muscle of her body becoming tightened and taut. After that, still barely conscious, with her body hyper-extended and straining, she continued to breathe low, rasping breaths that came every two seconds or so. She could not speak to us or reach out, but she was now back in her body and with us in the room.

Jeff had remained glued to his mother's frame. I reached down and encircled him in my arms, rocking him and caressing him. In time, he loosened his grip. "Jeff, you are holding her here. I know this is very, very hard. But she knows, and you know, that it's time for her to go on. Sometimes we just have to let go."

"But, I never told her I loved her," he wailed.

"Well, maybe you could tell her now," I answered. Jeff turned to face his mom and began to cry again. "I love you, Mom. I want you to come see me play football this fall!"

Joan's mother had come around the bed and she

wrapped Jeff in her arms. He was physically much less tense, and he was no longer playing that he was tough and unfeeling. He turned to welcome his grandmother's embrace.

"You know," his grandmother said, caressing his hair and holding his head on her shoulder, "your mom is pretty tough. And the way that she's loved you, and the way she still loves you, well, something tells me that somehow, in some way, she's going to be around, watching over you, taking care of you, *especially* when you're playing football!"

Tentatively reassured, Jeff turned back to the bed and took his mother's hand in his. "Well, I guess we'll be okay, Mom. But we'll miss you." He kissed her hand and then placed it carefully over her chest. With that gesture, Joan's physical body eased. You could sense her spirit issue a sigh of relief. She took several very deep, sighing breaths, and then one final deep inhalation in, and one final, but peaceful, breath out. And, she was gone.

The family stayed in the room for a while, touching and holding Joan's body, talking about what they had to do, who they needed to call, and where they should go. There were people to notify and arrangements to be made. A friend of the family had come to the hospital to drive them home, and when they were ready, I walked them out to the front door of the hospital. Stunned and grieving, they carried several small plastic bags of their mom's personal belongings to the car, moving in slow motion. I repeated several times what I was going to do and when I would see them again. They eventually understood, and then collapsing into the car in a numbed heap, they were driven home.

I returned to Joan's hospital room, where they had not removed her body yet, and where she lay, peaceful and very still. I came back to say my own goodbyes, which I felt would be best said in private after the family had gone. I had always been somewhat in awe of her strength and her courage, knowing all she had been through in her life, and she had become a close friend in a short amount of time. It was, to me, a real gift and an honor that she had asked me to be with her at the time of her death.

I looked at her body lying on the bed, certainly lifeless, soon to be cold. But beyond the area of the bed, above it and maybe hovering just below the ceiling of the room, over towards the corner, I experienced the very strong sense of someone's presence. "Joan," I said into the air, thinking of the Eastern and Native American traditions that teach that a person's spirit stays near the body for three or four days. "Joan, I, too, release you to go in peace. I know that this was a very hard thing for you to do. But you did a terrific job! I am so proud of you!"

The energy level in the room began to increase, and I felt something like a gust of wind, or a sweeping rush surround me. It was a strange sensation, something in the air, as if traveling from her molecules to my molecules, something imperceptible that came in waves of vibration from the upper corner of the room. It was as if words were being sent and I was receiving them on a different but very subtle kind of radio.

Then, somehow, from the other side, I could hear her laughing. I could hear her words. "Thanks," she was saying, "I *am* now finally at peace. It was a rough ride. But *never* underestimate the power of that peace prayer!" Her warmth and radiance filled the room as she embraced me and held me one last time. Then, finally, the energy level in the room waned, and the presence was gone.

I turned to walk from the room. Remembering for a moment all that had taken place, I closed the door softly and respectfully behind me. I paused for a moment to give thanks, to acknowledge that this had been sacred space, and that I had been invited to walk on holy ground.

Then I went home and put myself to bed.

❖ ❖ ❖

The intimate connection between music and death, sound and dying, as witnessed in this chapter, has a long and rich history. Although we in the West may not automatically think of music or sound on the one hand, and death on the other, there is something deeply ingrained within the human psyche that connects the two:

> The idea of the acoustic nature of the soul, which survived into the Middle Ages in Europe (*symphonalis est anima* said Hildegard von Bingen) and was taken up again by the German Romantics (notably by Friedrich Schlegel), is manifest among [traditional] peoples particularly in the belief that, even after the disappearance of the last mortal remains, the soul of a dead man (that is, the substance of the human being) survives as an essence (spirit) which is perceptible only in sound (Schneider, *New*, 10).

This belief in the soul as consisting of acoustic substance is witnessed worldwide in the universal use of sound in the form of prayers and/or songs during the religious rites for those who are dying as well as for those who mourn. "Song provides a road for the departed soul to travel upon" (Hale, 104). Special prayers and songs meant to give aid during the time of the dying transition are found in every major world religion. A devout Jew may wish to die with the sounds of the *Shema* on his or her lips, "Hear, O Israel, the Lord is one God, the Lord is One," while in the *Kaddish*, the next of kin will recite the mourner's prayer for a year following the death. At the time of death, a devout Muslim may wish to hear "appropriate *suras* from the Qur'an, [. . .] With the holy name of *Allah* in his heart, the Muslim is ready to pass into the next world" (Kramer, 164).

From the Buddhist tradition comes the practice of reading to the dying person from *The Tibetan Book of the Dead*, beginning at a time before the person has died and continuing after. For centuries, Hinduism has employed *mantra* in its rites for aiding first the dying and then the grieving. "Indomitable faith combined with supreme serenity of mind are indispensable at the moment of death" (Evola, 195), and *mantra*, or repeated sound, is used to grant such serenity of mind. The Sanskrit word *mantra* consists of the elements *man* (connected with "mind") and *tra* ("to protect," or "to cross over"). The syllable *man* is found in "standard words for insight and insightful prayer, [and] contemplation of the face-to-face relationship with god" (Findly, 43), and foundations for its use as preparation for death, as well as being a calming presence while dying, are found in the ancient texts of the *Rig Veda*, where it is mentioned that *mantra's* purpose is "'to ensure certainty' (RV I.31.13) and to 'bring joy on the most distant day' (RV X.95.1d)" (*ibid.*).

The importance of what a person is thinking, or in what sounds a dying person is engaged at the time of death, is emphasized by many spiritual traditions. St. Augustine wrote: "He who sings prays twice," as if to say that singing has double the efficacy of prayer. One of the primary purposes of prayer is to still the mind and quiet the senses. While tomes have been written concerning prayer, the world's great spiritual teachers seem to agree in recommending the use of short, repeated prayers as being most effective. In the sixth century, St. Benedict directed the monks under his care to read John Cassian, a fourth-century sage, for advice. "Cassian recommended anyone who wanted to learn to pray, and to pray continually, to take a single short verse and just to repeat this verse over and over again" (Main, 9). The purpose of such a short verse is to still the mind, enabling the person praying to first become quiet and then be able to listen for what God might be saying. "A thousand years after Cassian, the

English author of *The Cloud of Unknowing* recommends the repetition of a little word: 'We must pray in the height, depth, length, and breadth of our spirit, [he says] not in many words but in a little word'" (*ibid.*, 10).

In contemplative prayer, the focus of attention changes, new ways of perceiving emerge, and the person praying becomes aware of how intimately connected she or he is to the universe. "In neurobiological language, in contemplation we quiet the babble of thoughts produced in the brain's left hemisphere in order to spend time in the right hemisphere. Through this shift of brain activity, we open not merely our conscious mind but also our unconscious to the transforming power of God" (Roth, 73). Prayer (spoken, unspoken or sung) can profoundly affect the mind as well as the body, and has been incorporated since time immemorial into the religious rites of the dying. The sound of prayer builds "a bridge between the inner and outer worlds [...] and the power of faith is amplified" (Campbell, Roar, 70).

In this chapter, the fashioning of a "breath prayer," a short, repeated prayer like that of a *mantra*, comes from the usage described in Ron del Bene's book, *Into the Light*. It follows the method that he has developed to help people who are dying create a calmed presence within. As a hospice chaplain, I have found that breath prayers are a very effective way to bring comfort to those who are suffering from either physical or mental distress. (As a hospice chaplain, I have found that breath prayers are a very effective way to bring comfort to those who are suffering from either physical or mental distress.)

The use of sound has been proven to alleviate pain in many situations. People often make sounds when they stub a toe, or accidentally hit a finger with a hammer. Unconsciously, they know that the vibrations will somehow bathe the painful area with soothing stimuli. In *the Mozart Effect*, Don Campbell speaks about the many ways in which clients have used sound to soothe pain, including sending sung tones into the organs of the body. (Campbell, Mozart, 94). Dr. Ralph Spingte's work as an anesthesiologist has shown that surgery patients who listen to self-selected music during surgery need up to fifty percent less anesthesia during procedures, which results in safer surgeries and faster recovery times (Spintge, 82-97).

The soothing use of Gregorian chant is also mentioned in this chapter. Recently there has been an amazing upsurge of interest in Gregorian chant, as witnessed by the *Chant* recordings released recently by the monks of Santiago de Compostela, recordings that have soared to the top of consumer charts. This popularity has come as a pleasant surprise to theologians and musicologists alike, since the melodies of Gregorian chant, which in this century were still barely alive in the liturgies of the Roman Catholic Church, were all but consigned to the waste bin of history after the reforming efforts of Vatican II.

In reality, Gregorian chant is not only music. It is a "form of prayer, [and] cannot be understood merely through its music but [. . .] through the experience of prayer. It stands midway between the spoken word and pure mystical contemplation" (Schneider, "On Gregorian," 3). It is an avenue of connection with the divine, and in much the same way that *mantra* is used in the East, Gregorian chant is a gateway, "a route or path, a vehicle of movement. In pre-Christian symbolism it might have been considered a cart, a ship, or a river carrying clear sound syllables" (*ibid.*).

Gregorian chant is music that was originally composed to accompany and enhance worship; it is music with the sacred purpose of connecting the devoted more deeply to God. As such, rhythm and accent play an important part in that within the chant there is often a slowly moving, measured accent that coordinates with and even directs the breath. Balanced with the words and the meaning of the texts, there are groups of notes "so well balanced as to convey to, and produce in, the mind a sense of order in the midst of variety" (Benedictines, *Liber Usualis*, ix).

This "sense of order in the midst of variety" is perhaps what is currently being perceived by so many who now use Gregorian chant as a means for meditation or contemplation outside of formal worship. "It beckons us to sit down and rest, to enjoy not only the lovely melodies but also the silence from which they come and to which they regularly return" (LeMee, 4). There is a devotion to the sacred and an ability to rest in silence that is immediately discernible in Gregorian chant. Whether we can understand the words or not, we can perceive the loving flow and intent of the melodies. The slow, regular pulse of the music causes us to repeat its phrases and to copy within ourselves the music's deeper rhythms.

Perhaps people are also hearing again something in this music which medieval people knew. Katharine LeMee writes that "medieval people were well aware of the formative power of music. [. . .] They also knew that sound is causal, that it can bring about changes in the very nature and fabric of society as well as within the individual" (*ibid.*). The music theory of the Middle Ages reached far beyond the mathematics of sound to include an entire view of the universe that was musical. This musical paradigm, which medieval thinkers had found in their Greek sources, traceable to ancient models of Babylonia, Sumer, Egypt, and India was

> able to enshrine the knowledge of the universe in a very compact form, and a song or hymn built on harmonic principles embodies this knowledge. [. . .] As played on the instrument of the heart, mind, and body of the performer, the song's knowledge becomes a living reality. In this way the singers become a medium between heaven and earth, . . . bringing their listeners into direct contact with higher worlds. In that moment, living in accord with the laws of heaven and

earth, they resolve the dissonances within themselves and within their audience, spreading healing through their song (*ibid.*, 26).

It is, perhaps, this healing paradigm of the universal music model, with its harmonic and integrative effects, that so many people are intuitively responding to now as they listen to the beauty of the deeply flowing strains of Gregorian chant.

We have moved a long way from the belief that the "soul gains immortality by picking roses in the grove of the Muses" (sixth-century B.C.E. Greek poetess Sappho), that the soul consists of acoustic material, or that sound, in the form of prayer and song, can aid the souls of the dying. However, with the recent recurrence of the popularity of chant, we may be recapturing what we have known for centuries, which is that sound, with its soothing and therapeutic qualities, is our friend. Sound is our companion on the way, and when the time comes to make the transition from this life to the next, prayer and song and *mantra* can aid in the journey.

CHAPTER 7
BEING STILL–
"Don't You Miss the Music?"

After several years in the ministry, having worked with a few churches but having primarily served at hospice, I returned to visit the professor of musicology who had been so formative, so supportive to my musical development in earlier years. It turned out to be a pivotal visit.

"Don't you miss the music?" he asked as we sat one evening after dinner, reminiscing. We had been discussing those magical summer nights of years past when musicians from around the world had gathered for musical conferences in the Berkshire Mountains of Massachusetts to play the music of Baroque Venice or Renaissance Florence on original instruments.

"I miss it very much," I answered wistfully, for I did miss the music, and recently, a kind of harmony had disappeared from my life. The busy-ness of even divine service could fill my days and consume my nights. The demands of my self and the expectations of others resulted in a kind of giving that reached beyond my reserves. I was beginning to feel dry and parched, and the quality of my service was becoming dull and strained. I started losing the music in my own life. During the rigors of seminary, I had put the music down, and under the pressures of professional performance, I had even sold my instrument. Though grounded in Scripture, I had not been singing. And since not singing, it seemed there had been no real "word from the Lord" for some time.

Yet, it had been a "word from the Lord" that had drawn me into the ministry in the first place. I had heard a voice so ancient, so profound that it had made my heart stop. It had been, on the outside, the voice of an ordinary man, but it was, on the inside, the Voice of the One, who has been calling so lovingly all these years. The words had been like fire; they burned like ice and glistened like snow as they melted my heart into pools of water. It was in response to that voice that I had begun the years of seminary training marked by outward curriculum, while guided by the inward calling of a divine sound. It was in response to that voice that I had committed my life.

With one swift gesture, my professor, in his true perception and clear insight, had delivered the crucial question that stood calling in the center of my heart: Didn't I miss the music?

Yes, I desperately missed the music! And so, in the midst of the busy action of service, that still, small voice began to pull me from the constant motion of activity into the silence of meditation and into the depths of chanted prayer. I could feel that something was moving, something was trying to speak. Attuned to the question, grateful for the guidance, I determined to retrieve the music. I resolved right then that through the discipline of silence, I would seek to re-gather the sonic threads of my life, which somehow, unheard and unseen, had become unwound.

"How *silently*, how *silently* the wondrous gift is given" was the melody that floated lightly through my mind as I drove through the large, square pillars that served as the gateway through the old, ivy-covered stone walls of the cloistered community. Built at the turn of the century, the buildings had been constructed to look like architecture much older, more medieval. I drove my car into a parking spot, got my bags out of the trunk, and readied myself for a silent, reflective week of sorting things out.

During the evening meal, the week's schedule was explained, the rules of the community were given, and the assignment of a spiritual counselor was made. The overall rule of the community was that of silence. You were not to speak to people in the halls, or on the way to services, or even during meals. You could make phone calls during one hour in the evening. Visitors who were there on retreat could, of course, speak with their spiritual directors at the assigned times.

That first day of silence was very difficult. So many emotions vied for my attention that I found it hard to be quiet. Thoughts raced uncontrollably through my mind; they bombarded me with questions. The past incriminated with facts; the future teased with redemption. Pros and cons played tug of war, while memories and questions chased each other in circles. All this came tumbling out of my mouth at the first meeting with my spiritual director.

In a nonstop soliloquy, delivered at breakneck speed, I poured out my life, and all of my questions, in the space of forty-five minutes. "What happened? Where was the music? Which direction should I move in? When again would I hear God's word?" My spiritual director, having listened calmly and wordlessly, rounded out the fifty-minute mark that defined our time together with one very simple comment: "When you're not sure where you're going, it's a good idea to stand still."

That "standing still," I came to see, would take some time. The next day dawned with renewed fretting. I sat in my small room, alone and in prayer. Frenzied, I cried, "Lord, what would You have me do now?" The sound of the question overwhelmed my hearing, as the real work of inner quieting began. During the first couple of days, the discipline of silent meals helped to quiet the mind. This was not so easy at first, however, as I had been raised with the understanding that you were *supposed* to talk at the dinner table. Dinnertime was a time to engage people in interesting conversation, to inquire about people's work, or to ask how the day had gone. To sit quietly and acknowledge each person's presence through nods and glances seemed awkward and confining. Furthermore, if you needed anything, like the salt, you could not ask for it, as talking of any kind was not allowed. So, not wanting to nudge your neighbor, or point rudely, or engage in dramatic pantomime which would surely bring undue attention to yourself, you simply went without–all the while screaming on the inside: "Would someone *please* pass the *salt?*"

Gradually, however, by about the third day, those of us who did not usually eat in silence became somewhat more accustomed to the practice. We had begun to develop a kind of sixth sense, an awareness for each other's needs that was communicated on a non-verbal level, through a subtler kind of exchange. It was quite amazing what could be said without words and what one could hear without conversation. We were finding that in the silent process of watching for each other's needs, worlds full of words became extraneous and communication began to happen on a vibratory level.

These few gradual changes and the deeper sense of quiet that I was beginning to feel were topics for discussion at the next meeting with my spiritual director. I remarked that I felt my speech seemed to have become less frantic. There was more space between my words, more room to breathe, and more time to take in what I was hearing. I was speaking more slowly, and we were communicating on a deeper level, a level below the words.

My spiritual director nodded in agreement, acknowledging the inner quieting, and she gave me a verse that I was to contemplate and to repeat silently, as an aid to further development:

"Be still and know that I am God.
Be still and know that I am.
Be still and know.
Be still.
Be."

I fell asleep that night repeating the verse, "Be still and know. Be still. Be. Be still and know. Be still. Be."

The next day I wandered around the grounds. I wandered, with no specific purpose, from one spot on the grounds to another, repeating the chant in my head, "beee . . . beee . . . be still and beee."

I relaxed inwardly, chanting the sound, and found a comfortable, grassy spot on which to sit and observe what was going on around me. The ground had been warmed by the sun and the grass provided a soft cushion. The rays of sunlight penetrated beneath both layers of clothing I had on, taking the edge off the cool breezes that played on the wind. It was quiet there. I was quiet there.

"Real silence?" I mused. "Real silence, complete silence? That *would* be hard to come by." I listened to the "be" resounding within me, as well as all the sounds around. Although far from traffic and the noise of the city, the world was still full of noise. I could hear a squirrel or rabbit foraging in the underbrush; birds were calling back and forth insistently to each other. I watched as a bumblebee gently let himself down onto the petals of a flower, gruffly humming his own version of the "beee" chant. Chirping, humming, and scurrying were taking place all around me.

The wind was also singing that day. In the boughs of nearby pine trees, wisps of wind sang out "beee" in high-pitched phrases as they whistled through the long, thin needles. The wisps then curved to swirl themselves around a neighboring silver maple, emitting a lower sound "bhoo" as they encountered resistance in those broader, lower leaves. "Beee" cried the pine trees in a high-pitched whine. "Bhoo" answered the lower bass of the maples. The winds filled out their own harmonies, adding sounds and tunes as they danced among the trees and played among the bushes.

I sat, listening intently to the symphony that was sounding everywhere. Then, focusing inwards away from nature's sounds, it seemed as though "beee" was coming from my own body. My ears were hearing an inner sound and if I listened very closely, I could pick up a sound within or beneath all the other sounds. Closing my ears with my fingers, I could hear, deep inside of me, the sound of "bhoo" echoing and re-echoing in a low, fundamental tone.

I sat for a while, mesmerized by sound, contemplating all that I was hearing. My attention would shift from outer sounds to inner sounds, and then inner sounds would blend again to merge with the outer sounds. Like two sides of one coin, each was connected to the other. I remained listening intently for some time and finally wandered back to the main building for dinner in silence and evening prayer.

That night, as the gusts of wind against the windowpanes lulled me with the harmonies of wind-swept tunes, I fell asleep revisiting the counterpoint of those two songs. The higher, faster energies sang out the tune "beee." The lower, richer tones echoed back "bhoo." "Eee-eee" rang the one. "Ooo-oo" entoned the other.

Later, somewhere in the twilight between waking and sleeping, in the midst of that watery world one encounters between life and dream, I felt the touch of an indescribable presence. It was alive, but it was also like being in a certain place. It was all-encompassing and round. It pulsated, and like a net, it had many, many little connectors. Actually, it was more like a web, a web with many colors, made up of tiny, singing threads. The threads hummed as they vibrated back and forth, activated by little currents of activity that coursed along the strands.

Little by little, it became clear that two numinous presences were emerging. I could feel two very distinct entities delineating themselves from within this singing web. They both seemed to be female. The one left me with the feeling that she was Native American, and the other left me with the very distinct impression that she was from India. The two of them were singing and they proceeded to wrap themselves around me, encircling me. I could feel the humming of their beings enter my own, one from the left, and one from the right.

They didn't really speak, but they did communicate. By direct means, their thoughts became my thoughts, and I could hear what they were saying through the words of their songs.

They were singing a song about the beginning of the world and how the world continues through the singing of their songs. In one verse, they sang about being the ones who had sung the first songs. In another verse, they sang about being the ones who send singers to the earth, so singers can teach people the words, and people can sing the songs that keep life in balance and keep the world alive.

These two singers were very, very old. I could hear the age of their voices and I could fathom the depths in their resonant tones. They had sung the first songs, those that had created the world. They had spun the first energies, with words still coursing in multi-colored streams through the web of life. Theirs were the first sounds that brought silence into light.

Just as the two women were withdrawing themselves from my awareness, they gave me their names. They said they were known as the Song Weavers, or the Ancient Ones. Nothing could be older than they were, for all things had been sung by them into the web, and all things returned to them in time. Then they disappeared, humming, back to where they came. I could feel the sensations of the web vibrating in my body as I awoke, and then, gradually, all contact vanished. I lost touch with the sensations to find myself again awake and in the world.

"Who were those women?" I asked my spiritual director at our final meeting of that week. "And why did they give me their names and their song? How will I ever know what it means?"

"You'll know," she said. "Sometimes the answers we receive are not understood with total clarity in the immediate moment. But you'll know. The ones who resonate from that place of deep listening are always of great importance and you will come to know them, in time.

"You'll know, with time," she said reassuringly. As we said our good-byes, she repeated, "You'll know."

I closed the door to her office. Much more at peace than when we had first begun, I was buoyed up by a growing sense of expectation. The week of silent retreat had not been what I thought it would be. My attention to listening had shifted significantly and this recent "word from the Lord" was mysterious and new.

I packed my things, ending my retreat with a strong sense that my life was about to change in some unpredictable ways. I looked forward to the adventure, unsure of how it would manifest or what it might sound like. But, as I set out on the journey, in search of meaning and new worlds of sound, her words continued to ring in my ears: "You will know. You'll know."

※ ※ ※

Ultimately, this book is as much about silence as it is about sound, a topic addressed specifically in this chapter. Silence and sound are two halves of a whole in some philosophies, while they are one in the same in others. All sounds grow out of silence, while returning there as well, and yet there could be no grasping of silence without its manifested counterpart, sound. "Listening is active awareness of both worlds" (Campbell, Roar, 37).

In the West, the Word of God is very close to, if not synonymous with, silence. It was silence from which God spoke, and it is through silence that one actually connects with or hears God's Word. So, silence is considered to be "the attitude par excellence of the one who would hear the Word of God," and "Augustine [greatly developed] this in his teaching on return to the heart (*redire ad cor*), to the place 'where stillness reigns, where a true, spiritual seeing and hearing becomes possible'" (Gawronski, 114).

In the East as well, silence is the hidden substance behind, as well as the ultimate goal of, all sound. "Indian civilization, which, it seems, has more than any other [culture] given to speech or word (*vaac*) a central, basic role [. . .] has placed at the acme of speech, at the heart of every utterance, not sound, but silence" (Padoux in Alper, 297).

For the musician, sound and silence are partners. They are mirror images that sometimes reflect and sometimes play against, but always support, each other. To make sound, you must begin from silence; the musician takes time to focus, to collect her thoughts in order to begin. The exchange of sound with silence continues even during the simplest tune, with rests (or pauses) shaping and defining sonic material. And, when bringing the piece to a close, the musician marks the ending, returning the sound back to its silent source.

From the standpoint of human physical hearing, however, total and complete silence may never be truly attainable. The story of John Cage's entering a soundproof chamber in the hopes of obtaining total silence may be well known to some readers (Schafer, 256). Upon entering what was expected to be a perfectly quiet environment, Cage was still able to hear two things. He was still able to hear a high-pitched sound, which was his nervous system working, and he could hear a low sound, which was the sound of his blood moving and circulating. "Cage's conclusion: 'There is no such thing as silence. Something is always happening that makes a sound'" (*ibid.*).

Far from the city, away from traffic noise or the humming of electric lines and the roaring of jet planes, the country soundscape will offer up the sounds of a rushing brook, a buzzing bee, a calling bird, and the whistling wind. Beyond that, our own bodies will offer up whole "biological symphonies of sound" found in the "internal pulsations, the rhythmic beating and pumping of our hearts" that are the very noises of our being alive (Bush, 71). Life is sound; true silence would mean physical death.

Philosophers and theologians have characterized the twentieth century as an age of anxiety, an age uncomfortable with itself. To experience the reality of silence may have always been a rarity, but it is most certainly a rarity these days, and it would seem that the frantic rush of life avoids silence at all cost. We blast the radio; we turn on the television; we ignore the encroaching noise of traffic and the deafening roar of jet planes over our heads. Equating solitude with stillness or lack of movement, "[m]an likes to make sounds to remind himself that he is not alone. [...] Modern man fears death as none before him; he avoids silence to nourish his fantasy of perpetual life" (Schafer, 256).

And yet, what we fear is the very thing we need most. In an address to the first International Conference on Acoustic Ecology, Ursula Franklin wrote that "silence is an enabling condition, that opens up the possibility of unprogrammed, unplanned, and unprogrammable happenings" (Franklin, 6). Unlike religious rituals that presume given results within their performed actions, silence is offered as the commodity needed to correct current imbalances. As technological advances have limited the soundscape, the landscape, the mindscape, Franklin argues that "allowing openness to the unplannable, to the unprogrammed, is the core of strength of silence. It is also the core of our sanity, not only individually, but collectively" (*ibid.*).

Max Picard has also written about the twentieth century's sense of alienation, which results from being out of touch with silence. Human beings have a need for silence, as much as they have a need for food, for sleep, for oxygen to breathe. "Therefore, the longing for silence is ignored only at the cost of neurosis, psychosis, spiritual starvation and unhealthy human relationships. For any human society to ignore its need for silence is to point a gun at its own head and to slowly, slowly pull the trigger. Sooner or later, the gun will go off" (Finley, 4).

Picard felt that our connection to silence was essentially our connection to God's love. It is, thus, through the connection with silence that each and every human being can translate God's love into the world, being a channel of love for others. With a healthy relationship to silence, each individual can speak that love to others. Picard's greatest fear was that we, as a society, have now grown so out of touch with the health-generating love of silence that we have become sick and unbalanced from the lack of quiet. Quoting Kirkegaard from the previous century, Picard wrote: "'The present state of the world and the whole of life is diseased. If I were a doctor and were asked for my advice, I should reply: Create silence!'" (*ibid.*)

In many spiritual traditions, it is actually sound itself that directs you into silence. When using *mantra*, or repeated sound, the vibrations of the sound will calm the mind's inner dialogue and take one into a space of quietness. To quiet the mind in meditation and obtain a sense of balanced peace, one does not have to retreat from the city. "It is not necessary to give up the world and go to the forests. The stillness is the real seclusion" (Singh, S., *IV*, 19). The change can come simply by turning within, by being willing to quiet the mind and to sit with the silence.

We find in contemplation that we are not alone. For "when the mind is stilled, we become aware of the divine presence, the Lord of Love, who is enshrined in the very depths of our consciousness" (Easwaran, 81). This presence, in turn, connects us with everything else that is in existence. Far from being isolated and alone, we find, in the depths of our being, that we are deeply and fundamentally connected. Far from being a selfish, isolated act, the ability to quiet one's mind and to "find the still point within" is an act of community. For, in the stillness, we find that "all life is one. When we have had this experience, we will be incapable of doing anything that violates this unity of life, and we will live for the welfare of all" (*ibid.*).

Murray Schafer reminds us that this is precisely why it

is so important to our "spiritual metabolism" that we regain our relationship with silence (Schafer, 253). In past decades, society honored the library, the church, the shady grove—even the bedroom—as quiet sanctuaries where the individual soul could be rejuvenated in the solitude found therein. Specific spaces and times were set aside for the purpose of re-fueling the individual, in order to better serve the society as a whole. Before holy days became "holidays," entire days of the year were set aside for contemplation and re-balancing. "In North America, Sunday was the quietest day before it became Fun-day. The importance of those quiet groves and times far transcended the particular purposes to which they were put. We can comprehend this clearly only now that we have lost them" (*ibid.*).

A number of voices call us to re-establish our lost relationship with silence. I am indebted in this chapter to Patricia Hart Clifford's *Sitting Still*, a work in which she encounters silence. I am further indebted, however, to a teacher of my own who knew that silence is also inextricably bound to sound. The voice of the professor heard here in this seventh story proves to be that of a modern-day shaman as he asks the question, "Don't you miss the music?" The question was directed to me at a time when I was in a state of personal dis-harmony, and it sent me on a quest in search of the deeper meanings of sound.

The fact that this story stands midway between twelve stories, as a kind of fulcrum, is no mistake. For the professor asks what I think is a crucial question for all of us, echoing what has been asked of indigenous peoples for centuries, as noted in *Song and Silence* (Hale, 130): "when dispirited people in an indigenous culture seek help from a shaman or a medicine-woman they are asked four questions:

1) when did you stop singing? 2) when did you stop dancing? 3) when did you stop being enchanted with stories? 4) when did you first start to feel uncomfortable with deep silence?"

CHAPTER 8
SHAMANIC JOURNEY

I began my investigations into various traditions of sacred sound in many different ways. I plunged into reading; I pestered experts. I enrolled in classes; I participated in weekend workshops and immersion trainings. While not all the information in the chapters that follow describe in documentary detail exactly what took place in each case, I have attempted to convey something of the profound events I experienced within my own spiritual world as I conducted these sonic inquiries. One such inquiry concerned the shamanic use of sound.

We began our weekend workshop in shamanic studies by looking at the word itself. We were told that the word "shaman" comes from the Tungus people of Siberia, and meant "one who is excited, or raised," or possibly "one who knows." The word referred to those who, through a practice of joining with the spirit world, could achieve power over spirits for the good of others, and often for the sake of those who had been overcome or harmed by the spirits of another world.

The word "shaman" has come to be used to describe any number of healers or medicine people around the world who practice various spiritual arts, but the classical sense of the term refers to those who fall within the worldview of the Siberian practitioners. This particular worldview, comprised of several layers, includes at least three primary realms. The first realm is this world, open to the comings and goings of ordinary human beings. The other two realms lie beyond ordinary reality in the Upper World and the Lower World; though once open to all, they are now only accessible to the shaman. This understanding of reality is one that sees the relationship between body and soul as more fluid than our Western view, in which the soul is encased within the body, and as such, the shaman's soul is able to leave the body and travel with ease among all three worlds.

At the center of the shaman's universe stretches a great World Tree. This Tree is the connector between the Lower World, the Upper World, and this world. While the shaman's drum is fashioned from a tree from this world, it signifies its cosmic counterpart, the World Tree, and the wooden rim of the drum functions to connect this world and the world of the spirits. The drum is often referred to as the shaman's "horse," since the shaman takes his or her celestial ride to other realms on the back of the sounds of the drum.

In these other worlds, accessed through sound, there are spirit helpers, often animal spirits, who help the shaman in his work. Contact with these spirit helpers is made easier since an animal's skin, specially sacrificed for this purpose, is stretched across the wooden rim of the drum. Within the branches of the Great Tree, it is said, there live birds that are people's souls. These soul-birds live in the Tree before being born into a body, and it is there that the soul-birds return when separated from the body by trauma or death.

Modern society, until very recently, has looked condescendingly upon the shaman as a "primitive" and superstitious being. However, the shaman's role in his society has always been that of healing sickness. By traveling to other worlds via the Great Tree, the shaman is able to retrieve parts of a person's soul that have become detached from the whole of his or her personality. The shaman is able to release people from pain and illness sent by the spirits. The shaman brings order to society. Through willingness to interact with the dangers of the spirit world, and by overturning the forces of evil, the shaman is a representative of healing and wholeness, a bringer of the Light.

Our initial experiments in journeying were surprising and magical. First, we had journeyed for another person, with the intent of bringing them a gift or a symbol for their own healing. Then, we journeyed to hear our secret names. Beneath the choppy sibilance of my own name, I had been in touch with smoother, more continuously flowing sounds. I had become aware of worlds within my own psyche reached through the agency of repeated sound.

Those worlds were personal, inner worlds, and they were very real. Yet, at the same time, they reached vastly beyond my own limited self. They were extremely old. Others in the group had shared similar sensations, and we came to the conclusion that we had somehow traveled together into these ancient regions. The drumming had taken us into areas of reality not included in the general awareness of our everyday life. Those areas, primordial and archetypal, may have once been familiar and open to all, but now they were avenues closed and secret from disuse. Through the agency of sound, we had stumbled once again onto those well-traveled psychic maps, now uncharted territory for the modern mind.

Our final assignment was to undertake a journey for ourselves. Armed with the healing symbol we had been given, and accompanied by the power generated in our secret name, we were to travel either to the Upper World or to the Lower World. There we would find for ourselves what we needed to know, or hear, or encounter that would bring healing and balance to our lives.

Just as in earlier journeys, we laid on the floor on our backs, in a very relaxed position, listening intently to the drumbeat executed by the workshop leader. The constancy of the reverberant drumming served as our means of transport for the journey, and we simply followed where it led. I relaxed into the mesmerizing sounds of the droning, repetitive beat. The current of sound filled my awareness and surrounded my body with deep, resonant tones.

The low cries of the animal skin began to emerge and the spirits of the drum were gradually released. Slowly, I began to climb, limb by limb, up the great and ancient Tree, as I began to float high above my body, far beyond the room, within the darkness of inner space. The Tree reached out in front of me as far as I could see, and I could feel the tactile impressions of its bark on the palms of my hands as I climbed, carefully and methodically, pulling myself up, first on the right and then on the left. Brushing past leaves that were centuries old, I wound my way upwards through the branches. I sensed my feet clutching each arm of the tree, now on the left, now on the right. I grasped each branch and pulled myself up, limb against limb.

I was beginning to feel that I was actually dissolving and becoming part of the Tree, when I felt powerful legs push off from beneath me and I was launched into the air. The continuous drumming penetrated my skin, repeating hypnotic pulsations, and creating the sensation that I was floating freely, hanging suspended high in the air. I merged into the sensation of feathered wings pulling vigorously through swirling eddies of cloud. Then, from somewhere inside an airborne, avian body, I gazed down to view the landscape below.

The land I viewed was very dry, having been split open long ago from the pressures of being baked in the sun. From the dry plain, great rock formations jutted upward, cragged and rugged. Descending on drafts that brought me closer to land, I could make out several high cliffs whose walls were draped with greenery. The lush vegetation increased as I swooped down to fly between two walls that wound round into a steep canyon. Gliding past mossy ledges and dark crevasses, I followed the walls as they curled to their conclusion, where I was surprised to find, sheltered in their midst, the rushing cascades of a waterfall.

The sounds of splashing water engulfed me as I flew nearer to the crystal stream. Liquid ropes of water sparkled in the sunlight as rainbows played in the sprays of mist spewed against the canyon walls. I plunged into the clear, purifying waters and allowed the splashing sounds to wash over and through me.

The tops of the cliffs then came into focus as I approached their summits from below. My wings strained against the strong, swirling currents, as I flew towards my goal of the mountain heights. In my talons I carried a large egg. The ecru egg itself was ancient, and just below its sandy-coated surface brimmed a fiery effervescence that glowed with the white heat of an electrical storm.

I could hear rhythmic ceremonial songs drifting down from the rocks of the cliff tops. Focused on the songs emerging from the rocks, I bore my way through the currents, guided by sounds through the wind. The melodies drifted down along the high-cliff walls, and the ancient, rhythmic cadence called me to a specific spot deep within the stones. There, into a nest, the egg was to be delivered, wrapped in the sound and delivered safely home.

As I headed into the strong wind currents that surrounded the top of the canyon walls, I was vaguely aware of how high I was flying. Far above the realms to which I had traveled before, I was completely out of reach of the normal world and entirely beyond communication with the group. I was floating freely in a separate world whose rules I did not know and whose geography was beautiful, but strange. I felt the measured release of iron talons opening gently with a delicate touch. The ancient egg, rushed through sounding currents, had been safely released into its mountain home. It lay now, protected, nestled in the cliffs of singing stone.

I was turning around to begin my journey out when my eye caught a glimpse of something directly behind me, gliding, shadowing, following me. It seemed that there was some kind of shimmering body or being, flying aloft, and it was coming towards me. A flash of light shot past my head and I turned to gaze into the face of the most beautiful, radiant being I had ever seen. Angelic and ethereal, it was translucent and numinous. Hovering in the air, streams of loving, flowing light radiated from its center and surrounded me. From its mouth came mellifluous strains of celestial harmony. From its eyes blazed the sensations of deep and sincere care. Frozen in mid-space, I was captivated by the warmth and beauty of such a bright and luminous form. Attracted to it by unseen forces, I reached out to embrace its light and engage its beauty when, as quickly as it had appeared, it disappeared.

Suddenly, from behind, I felt a tap on my shoulder. This time, I turned to watch as ethereal wisps of light evaporated into thin air. In their place hung a vacuum. Very different forces then coalesced, coming into sight. Here now was the energy of a being so dark, so cold, so heavy, that gazing into its eyes was to hear, echoing within oneself, the dank clanging of iron against iron as the prison cell's door closed for one's own execution.

I was caught in the death-grip of that ringing sound. Suspended and transfixed, I could not move. The beams of sparkling, loving light had become tendrils of vicious intent. The eyes, which had been so full of care, were dark and cavernous now, as they sent out arrows, sharpened like barbs of piercing poison. Small, poisoned darts left their dark sockets and wormed their way into my skin, piercing my flesh, spiraling deep within. The arrow shafts were long and wiry, and with the crack of a whip they lashed themselves around my frozen frame.

Shot with such force, arrows with their binding wires sang through the air–zing, zing, zing. Each long strand, taut and unyielding, held me paralyzed–the adder's poison, scorpion's tail, tarantula's venom, manta-ray's sting.

I could breathe now only with great difficulty. My lungs, my upper body, my lower limbs had all been bound. Then, out of nowhere came a shiny, flashing, blue-cold blade. Cunningly, silently, the surgical steel sliced through my skin before it could even register in my poor, numbed brain. The cut was not visible, hardly to be noticed, but a dark, red-brown liquid began to gather and ooze. I moved to look down, and there, gaping through the only place that had remained unbound, were my organs, my entrails, my vital life forces drowned.

I looked back over my shoulder to find the Great Tree, where, on one of its branches, a small, white, iridescent bird hopped back and forth, watching me. Though the bird was completely engaged in whistling its triumphant tune, I desperately called for it to come and help me. Still whistling, the bird flew its perch and spiraled down through the air towards me. Enfolding what was left of my entrails into its feathered, iridescent wings, it wrapped its body around my frame. As the heat of its body melted into mine, the cold, deadening tendrils shot from the body of the vicious being began to loosen their grip and fall away from me.

Finally, again, I was able to breathe. The frozen eyes released their hateful stare, and the creature of the cold faded back towards the icy dark. So relieved was I that the binding had ended and the damage assuaged that I hadn't noticed that all during this time the heat from the white bird's body had melted so deeply into mine that it had begun to burn and to immolate my skin.

I reached down to put out the fire, but my fingers were feathers, and my arms were wings. My inner organs burst into flame. Raging and ravenous tongues of fire devoured my entire body. From red-yellow to blue-orange, flames engulfed me, changing everything first to bright blinding light and then to black. All that remained was an outline of a crust, like a funeral pyre, charcoaled and ruined, ashen and black.

I watched then as out of the ashes a small gray wing began to rise. Pulling up and out of the ruins, it took shape with a song and rose from the pyre. Triumphant and strong, the melody expanded in the shape of a small white bird. In the folds of what had been organs were now nestled feathers, beak, and wings. Deep from the cavern resounded a melody with the words: "You cannot kill me, though you may try; light to the world, I shall not die."

The melody repeated again and again, soaring and falling, as the mighty bird, now placed deeply within my center, sang out its song.

"You cannot kill me, though you may try; light to the world, I shall not die." The pounding rhythms circled in on themselves, over and over, and over and over, only to return to the beginning to begin again.

"You cannot kill me, though you may try; light to the world, I shall not die." The drumming continued incessantly, unceasing. It raised itself, at last, into a frenzied crescendo, and then suddenly it stopped.

Through the expansion of reverberating silence, I gradually remembered that I had been journeying. My awareness exercised a slow-motion backwards somersault, and with a high-pitched sound registering something like "hhheeeeeeeee," I returned to the conference room.

Stunned, I looked down at the area of my solar plexus. From the outside, everything looked normal. There were no traces of arrows ingested, no tracks of long-tailed wires. There were no smoldering wisps of smoke; there were no indications of where I had been, and no white feathers melting into my skin. I inhaled deeply and exhaled with great force, hoping to clear myself of the experience. With the release of the out-breath, ghost-rhythms echoed as "Light to the world, I shall not die" shuddered through my bones and danced across the surface of my skin.

"Would anyone like to share their experience?" asked the instructor. I watched as a dozen enthusiastic hands shot into the air. I attempted to formulate a few brief impressions, but it was as if a screen had been placed between the experience and any attempt to describe it. I simply could not find the words.

We closed the meeting with a ceremonial circle intended to bring both healing and power. I stood open to accepting both, thankful indeed for spirit helpers and eternally grateful they had come to my aid. I hadn't known what to expect from the encounter, but I left acutely aware of the arenas of power.

I probably should have spoken about the mortal combat I experienced in my vision; it might have been better to let someone know. But I was focused on the Light Song sounding inside me and to speak too soon might have stunted its growth. Some things need silence in order to blossom, and I knew that in time, given half a chance, the Light Song was one I'd know.

❂ ❂ ❂

Following the seventh story, which considers sound in the context of deep silence, the stories move out into a retelling of experiences encountered while studying a number of different traditions that honor sound as a spiritual path. This story considers shamanism. As mentioned in the story, shamanism is a widespread phenomenon and its practices are many. In *The Shaman*, Vitebsky notes that shamans are found practicing in cultures worldwide; however, the "classical" definition of shamanism is that as found in the practices of such people as the Ostyak and Vogul peoples of western Siberia.

In shamanic cosmology, seven levels of the world are described, with three being accessible in some way. The Upper World is described as a world of gods and spirits, "where teachers and guides may be found, and journeys here may be particularly ecstatic" (Walsh, 147). The Lower World is described as a world inhabited by animals, a world where death lives, "a place of tests and challenges, [. . .] where power animals are acquired and the shaman is guided and empowered to victory" (*ibid.*). The middle world is the present world where we live, accessible in ordinary time to ordinary human beings. Only the shaman has the ability to access all three worlds through the practices of journeying.

According to Dr. Katalin Lazar, an expert on Siberian shamanism, there are two musical phenomena central to Siberian shamans and their journeying. The first is whistling. It is common for the shaman, while journeying in trance, to come into contact with the souls of the dead, who often appear as birds. The shaman will make whistling noises and various bird calls during trance to indicate that the spirits are present. The birds assist the shaman in flying from one world to another and the shamans refer to them as "my helpers."

The other constant is the drumming, which occurs continuously during trance. It is the drumming that creates the trance state, sustaining it as well, while the shaman journeys. The reason for the drumming's effectiveness is commonly believed to reside in the speed of the beat; if the drum beat is maintained at a rate of two hundred to two hundred and twenty beats per minute, it has been proven to induce trance (Gore, 8). The rapid drumming has the ability to transcend time and space (Eliade, 171) while it aids in focusing the attention. "Heightened concentration seems to be a key element in effective spiritual discipline, and shamans appear to have found one of the quickest and easiest ways to attain it" (Walsh, 174).

The drum is made from a wooden frame cut from a specially chosen tree, symbolic of the universal World Tree, and the skin of an animal whose life has been sacrificed for the specific purpose of becoming a sacred drumhead. "Siberian traditions say that the sound that becomes audible through the skin of the instrument is the voice of the being that was sacrificed to make the drum" (Schneider, "Acoustic," 70). The drum is often referred to as "the horse," for it is on the sounds of the drum that the shaman rides during his or her trance-state, and the drumstick is "the whip."

Dr. Lazar mentioned that for the Siberian shamans whom she has studied, drug usage was seldom found as a means to induce trance and that it was viewed as being suspect. Mircea Eliade also writes that, while drug-induced intoxication does occur, it is a rather late derivative in the practice and that it is considered to be an artificial means. Rouget makes the distinction that, while the shaman goes into a trance state, it is a controlled trance, one that the shaman orchestrates as he encounters the worlds of the spirits, and different from the possession trance, in which the person going into trance is taken over and controlled by the spirits (Rouget, 23). The drumming, rattling, and chanting provide the path for the shaman's journey.

The drum carries the shaman to the center of the universe, or the Cosmic World Tree. The World Tree is the central axis of the universe, and it joins all of the worlds—upper, lower, and middle. Climbing the Tree becomes a means for the shaman to enter into other realms, as the Tree is the bridge between the worlds. In the branches of the Tree sit many kinds of birds, perceived of as being the souls of the dead, and depending upon the situation, the shaman may enlist the aid of these souls to help him in his journey. Or, if one of these souls who has died is continuing to wander aimlessly or to affect the living in adverse ways, the shaman's job may be to escort that soul to the land of the dead, thus relieving pain and suffering in the land of the living.

There are two symbols found in this story that do not have specific reference to shamanism. One is the phoenix, a mythical bird; the second is the cosmic or world egg. First, the phoenix is an ancient symbol of re-birth, of the refusal to allow darkness to overcome the realm of light, and the affirmation that even out of destruction can come new beginnings. It is connected with the concept of self-renewing light, and "according to the *Haggadah*, the phoenix is a vast sunbird who 'spreads out its wings and catches the fiery rays of the sun.' On these wings are enormous letters, saying: 'Neither the earth produces me, nor the heavens, but only the wings of fire'" (Walker, 407). It is synonymous with new life born from the ruin and destruction of fire and ashes.

The cosmic egg is another *Logos* symbol of sonic creativity. At the beginning of creation, "Brahma [the Creator], in the form of the supreme swan, descends on the surface of the primordial ocean, to place the divine egg, *brahmaanda*, [. . .] the cosmic egg, which, upon opening, will give birth to the world and its becoming" (Filippi, 17). Like the harmonic reality found within the overtone series (wherein all notes can be derived from one), the world egg is a sonic creation of the All-in-One. "The Cosmic Egg of mystical iconography carried all Arabic numeral and alphabetical letters combined within an ellipse, to show that everything that can be numbered or named is contained within one form at the beginning" (Walker, 5).

Of all these chapters, which consider various aspects of sound, the eighth chapter brings up the question of music's darker side. It is brought up here in the context of shamanic practice, but it is in no way meant as a judgment or a negative statement concerning the shamanic arts. It is simply meant to say that in shamanic practice, it is quite possible to be in contact with the darker sides of both human and spiritual realities. The multi-layered cosmology of the shaman contains a wide variety of spirits, including those that may be malevolent or may be seeking to do harm. While the shaman leaves his or her body under controlled circumstances, the "mastery of spirits remains highly precarious" and there is a "constant risk of insanity or death" (*ibid.*, 11), which is physically and psychically very dangerous for the practicing shaman. "Healing the victim of a sorcerer may involve doing battle with the aggressor as well as saving the patient. In addition to darts and harmful objects, sorcerers may send their familiar spirits to attack a victim or even eat the victim's soul" (Vitebsky, 74).

As musician-priest, the shaman has the ability to enter worlds that others cannot in order to be in touch with those spirits that have the power to harm or even to kidnap people's souls. It is the shaman who has the power to meet and defuse the potency of those spirits, to guide them to resting places, and to bring relief to patients who have been plagued. The practices of shamanism are currently receiving a great deal of attention from Western, mainstream culture, in the form of experiential and experimental workshops and trainings. Chapter 8 is a gentle reminder that these practices are genuinely very potent and that they need to be approached with the greatest respect and care, honoring them in a deeper way than can be fathomed in a couple of weekends.

Chapter 8 also points to the fact that shamanic practices can be profoundly healing. "According to psychiatrist Erich Fromm, the basic human problem is 'how to overcome separateness, how to achieve union, how to transcend one's individual life and find at-one-ment,'" and shamanism provides a shift in perspective that transforms a view of separation and alienation into one of unity and interconnection (Walsh, 259). For the shaman is a bringer of light, a healer–a savior, of sorts. The worldview of the shaman (accessed through sound) offers to us what is ultimately an alternative way, a musical way, a way that perceives of all things in terms of harmony. "For the shaman all is sacred and alive, everything is interconnected and interdependent, all creatures are part of one great web of life that holds all things in harmony" (*ibid.*, 255).

CHAPTER 9
THE LISTENING LODGE

We had gathered this time inside a beautiful, elegantly appointed retreat center serenely situated in the deep woods. Tall glass windows, accented by huge, rough-hewn wooden beams, gave a full view of trees and sky. It felt as though we were sitting in the midst of Nature herself. Outside, the setting was breathtaking and inspiring. Inside, however, we had reached an impasse. The only conclusion I could come to was that we were not really hearing one another, at least not on a level that counted.

Previous dialogues here had centered on the Native American understanding of sound and song, and the conversations with several Native American elders from different tribes had been profound and wonderful. Then, somehow, the suggestion of conducting a sweat lodge to experience sacred song in a ritual context had been introduced into the conversation. Now, as we met after dinner on a late-autumn evening to begin preparations for that sacred ceremony, it was as if we were caught in an invisible barrier we could not push through. There seemed to be some unvoiced disagreement about whether we should proceed with the sweat lodge on the next day, or not. Inside, we had reached an impasse–outside, it had begun to snow.

Small, crisp flakes of snow glanced briskly off the great glass sheets of windowpane, creating a light-filled symphony of tiny, prickling sound. As I listened to the crisp flakes hitting hard against the glass, mingled with the crackling hiss and spitting flame of the early-evening fire, the animated sound-play recalled our previous conversations.

The dialogue had focused on a worldwide web–not one of computers, but one of sound. We had spoken of the web of relationship that wove together all of Creation. This was a sounding web, as the whole of creation sang, and each member had its own song. The rocks, the trees, the birds, the insects, the people—all sang their songs, and the songs were woven together in order to keep creation alive. While chanting songs, the singers united earth and sky. In joining one's self to the Creator, and having received the gift of songs, the singer offered back the gift of sounds.

We had spoken about songs as medicine. Like wild fruits, plants, and herbs, songs, too, were given for healing. Some songs represented a whole tribe; others were given to an individual to be used for healing—healing of the self, or healing of others. Medicine songs could come from many places. They could come from the rocks or the trees. They could be given at birth. They might come from an animal; an animal could bring you a healing song. They might come during sleep, given by the Creator in a dream. Or, they might come during a sacred ceremony—a song might be given during a sweat lodge.

We had spoken of how we *were* the rocks, the wind, the trees. We *were* each other, all related through the sounds and through the Creator. Listening, we could hear the sounds that come from the trees, from the wind, and from the rocks. The songs were not separate from us, yet we had to be open to their coming. It was as if each of us was a television set or a radio, able to receive information on certain channels. If we were open enough, sensitive enough to many channels, we could hear things from all corners of the world and all ages of the universe. However, most of us were programmed to tune in just a few stations.

An analogy had been offered to illustrate the difference between an all-wave radio open to every band wave of AM and FM, long and short, and an ordinary, kitchen-model table-radio so common in the 1950s, capable of only picking up signals found between 16 and 1560 hertz on the AM band. To hear the songs, one had to be like an all-wave, omni-directional radio. Songs could come from the past;

songs could come from the future. They could be in any language; they could be from another time. But, when it came to listening, most of us used our ears–and lived our lives–like common kitchen radios.

One of my deepest, unvoiced prayers to the Creator had been to receive a song. Disappointed by the uncertain fate of tomorrow's ceremony and its lost possibility of receiving a song, I reached past the metal hanging curtains of the large fireplace with a pair of fire tongs to re-adjust the half-charcoaled remains of the wood, scrape together the red-hot embers, and place another log on the fire. Yellow and blue tongues of flame instantly attacked the dry edges of the bark, while little orange fire spirits hissed their way into the centers of the logs.

Outside, the rate of snowfall was increasing, and the white stuff itself was becoming wet and heavy. The wind was beginning to pick up. While listening to its haunting sounds, we shifted our attention more deeply within. Settling into the warmth, we moved into those stories that directly concerned the present. In a nearby Native American community, there had been abuses reported concerning the use of sacred ceremonies. A low, moaning sound wrapped itself around the building. Singers, who should have been purifying themselves in preparation for their sacred roles, had come to the ceremonies early and had sung their songs while being drunk. Under the influence, they had sung songs that were of the most sacred character completely outside of their ritual context simply for the entertainment of white audiences. The wind let out a sudden, high-pitched shriek as it rounded up one of the corners of the building, slammed itself against the wall, and burst through a tiny gap in the insulation, as a cold, searing presence swooped down into the room.

The light from the dancing flames cast long, strangely shaped shadows out of the fireplace, engulfing us in dark figures that danced in the fire light, masquerading as hands, fingers, or claws. During those same ceremonies, it was reported, other very inauspicious events had taken place. A sacred eagle-bone whistle had been stolen; people had become ill while dancing. One family had experienced an auto accident on the way home, and many other stories had been shared concerning bad luck and misfortune experienced by those Native people who had attended that event.

Very real concerns were being expressed. Traditions had been broken and ancient taboos had been transgressed. Some sweat lodges were allowing white people to participate now, and, as if they were out for Saturday-afternoon entertainment, they were being charged exorbitant fees by certain Native "holy men." Recently, even women had been allowed in the lodge. The concern expressed regarding all the misuse and disrespect for tradition was that additional misfortune could surely be the only outcome.

Outside, shapes moved in the night. Tree branches sagged and then rebounded under the weight of invisible beings. Plays of light and shadow showed faces moving among the trees. Gusts of wind hurled themselves at the great glass windowpanes, sending shudders through the room and into our hearts. Dark shapes lurked cat-like across the landscape, prowling impatiently, then settling down to watch us as we talked.

One of the Native elders rose to stoke the fire. Pausing for a moment to gaze into the center of the heat, listening to the sizzling drone of flame and enjoying the waves of warmth being radiated, he returned then, in a slow, measured fashion to his seat. Having completely taken in, and considered in its entirety, all that had been said previously, he began to speak. He agreed that there may have been abuses, and that some of them may have brought injury to those involved. He agreed that traditions needed to be honored, for only by staying on the path could balance be maintained. He agreed that caution was important. However, he said, he had also come to see, over time, that many things which he would have thought impossible at one point were, in fact, quite possible when seen in a new light, when viewed through the eyes of Spirit.

He had actually only come here, he relayed, because of a dream. The Creator had recently given him a dream of a sweat lodge, a sweat lodge that was to be. Normally, he would not have spoken of a dream until it had come true, but he felt that this dream had been about expanding the boundaries of what was considered conventionally possible, so he wanted to share that dream at this time.

The sweat lodge in the dream was situated on a plateau in the clouds near the top of a mountain beside the waters of a deep blue lake. It was of tremendous size. Willows had been chosen to create its ribs, as willows often grow near rivers and have a special relationship to water's ability to cleanse and energize. Saplings had been brought from all over the world to shape its dome. Animals from all continents had given themselves to furnish its coverings. Each element of creation was represented in this sweat lodge– the plant world, the animal world, the stones, the birds, the air, the water, and the fire. A microcosm of creation; all were present to create this lodge.

The rocks for the fire had also been brought from every land. Round, igneous rocks, they would hold heat for a long time and not crumble when water was poured onto their burning surface. All life had originally begun with the rising of a great stone from the waters of creation, and during sweat lodge ceremonies the rocks were heated to very, very high temperatures. When water was poured on the heated stones, there was a return to the dawn of creation, and the hissing of the steam that arose was the whispering of the words first spoken by the earth at the time of its birth.

This lodge, like all lodges, was meant for healing, but this lodge was to bring a new kind of medicine. This lodge, like all lodges, took one back to the beginning of time to listen to hissing stones, and thus be reborn. But this lodge was meant to bring complete rebirth; this lodge was meant to regenerate all the nations. This lodge was for all the peoples; this lodge was open to all who would come.

Four passageways led into this great lodge, marking the four directions of the earth. At each of the cardinal points, an animal image could be seen. Connected through vibrating energy lines, the images of the animals hovered over the lodge: to the north was a large black bear, to the west was a yellow coyote, to the south was a red fox, and to the east was a white wolf.

Within this lodge, the songs of all these animals mingled together. The songs of the black bear were chanted along with songs usually heard only from the coyote. The chants from the red fox harmonized with those from the white wolf. The songs, which were fervent prayers, all wound together. Inundated in the hissing clouds, they rose in swirls of steam, carried into the ears of the Creator. The prayers were sent for forgiveness, for healing among all peoples; the songs were sent for the healing of all the earth.

Outside, I noticed that the winds had died down. A stillness in the air could be perceived and the snow that had fallen earlier lay draped peacefully across everything like a warm coverlet. Shadows no longer lunged across the landscape and the tree branches were quieted. Inside, the fire burned steadily, casting a warm and welcoming glow.

We sat for a long time in silence. The only sounds were that of a low hissing and an occasional crackling from the fire. We sat in silence, wandering. We sat in silence, drinking in the beauty of the dream.

Silence is said to be golden. The issues that had been raised were important ones. Silence is profound. Traditions honored mean duties upheld. Silence is wondrous. Grasping a new vision of freedom, revealed through risk taking, could bring a deeper, long-lasting peace. However, silence might mean a stalemate. The dilemma was an old one. Silence could mean a standoff. Unable to move, we were caught between the spirit of what had been and the spirit of what might be. Silence had become a draw.

After some time, the oldest member of the Native Americans present cleared his throat. Deliberately drawing in a full and measured breath, he began to speak. He had originally brought us the analogy of the kitchen radio, and as he summarized the evening's events, he raised that image up again. Things were not always, he remarked, what they seemed. What seems closed today could easily be open tomorrow. What we couldn't hear today, perhaps we would hear tomorrow. Barriers that seem to exist between the past and the present are not etched in stone. Boundaries that shape a difference in the present and the future are only an illusion. There really is no difference between space and time. All are joined in the present moment, which itself gives only the illusion of continuing to undergo change. He encouraged us to stay open to the future. He pleaded with us not to give up on the word "possibility." Then, bringing his words to a close, he asked us to look forward to the time of a brand-new day.

The embers of the fire had all died down. Tired and exhausted by the evening's intensity, no new logs were being offered to the flame. Knowing we would get no further, we each excused ourselves respectfully, and, one by one, went to bed for the night.

That night I had a dream. In the dream, I stood gazing out onto a great plain. Expansive fields of tall grass surrendered in graceful patterns to the fingers of wind as the long stalks were caressed in airy, undulating sequences. I was gazing out onto the vastness of the plains, and up into the grand expanse of the arching sky, when I heard– or, more accurately, felt–the low bass rumbling of a great herd of animals moving over the land. The sound changed then into the heavy cadence of a single animal's great, loping frame.

The pounding of the heavy footsteps came closer and closer, and as it resounded in my ears, I looked up to find myself staring directly into the massive face of a great buffalo. Less than four yards from me, it had come to a stop. Fixing its gaze on mine, and staring into my eyes, he focused his energy on me. We were locked in encounter. I had never seen anything so huge! Its massive head, adorned with horns yellowed with age, hung suspended from its behemoth frame.

The rest of the animal's body was covered in brown-black fur, matted with dirt, encrusted in mud. The odors rising from his fur were strong and wild. The animal's breath was steaming hot, white as it hit the cooler air of the plains, and I realized how powerful this magnificent beast truly was. It stood staring at me from one eye and then began to move slowly, deliberately towards me.

Fixed in its gaze, I cautiously attempted to back away from the great animal, but somehow I lost my footing and crumpled backwards onto the ground. The great buffalo continued to advance purposefully towards me. I watched as he took that final step which brought his massive frame directly down onto my chest, directly over my heart.

"This ought to be killing me," I thought, as I watched the buffalo moving down into my rib cage. I inhaled deeply into the sensation of the buffalo's hoof piercing my heart, and released the fear that had gripped me, creating an inner space that allowed this greater Being to fill my lesser form. Exhaling, I embraced the experience. The buffalo then reversed its gait and magically retracted his leg and hoof from my chest. Miraculously, I was still alive. My ribs were actually intact. I was still breathing.

The buffalo looked once more directly and deeply into my eyes. His gaze riveted me in its light. Delivering a very specific message, I heard the buffalo issue one word. It was simply: "Listen." The message was "listen." He then turned away, and with the same lumbering, measured gait, shaking the whole earth as he walked, returned to the plain.

I awoke, startled. Half awake, I fumbled out of bed to raise the window for some air. The night air was fresh, and the moon's light shone down softly through the patterns of lightly falling snow. Small gusts of wind moved through the pine trees, caressing the deep green velvet needles. I breathed in deeply, asking the sharpness of the air to clear my head and cleanse my frame. Gradually, my breathing calmed, and the word "listen" seemed to repeat itself with each breath taken in.

Occasionally, a high-pitched sound moved through the upper branches, spiraling down around the trees. Or a low, resonant vibration would filter up to play among the lower limbs. The winds were lulling me, clearing me. As I stood lost in the music of nature, I was overcome with joy. Inspired by an awareness between bliss and ecstasy, every hair on my body stood on end as the spirits of the wind came into me. And, in that moment, I heard a song.

A melody drifted down, materializing out of the dark night sky. From some other place, it entered on a breeze, floating toward me like a ribbon of wind. I felt it through my breath; I heard it through the pores of my skin. The sounds were old, spiced in the flavor of a Celtic tune:

O Marion, yours was the sorrow.
Yours was the pain.
You will not know until tomorrow
What it is that you have gained.

I paused, repeating back the verse, first the melody and then the words, again and again, so that in the morning I could remember and be able to repeat the song.

I had a certain feeling about what would happen tomorrow. Overnight, we would get more snow. By morning, more than two feet would cover the ground. Without the proper preparation, there would be no ceremony. Without the proper agreement, there would be no sweat lodge.

But I realized, beyond my disappointment, that something else had actually happened. Equally wonderful, another kind of lodge had taken place. We had already participated in a type of worship, as we had all celebrated in a "Listening Lodge." And I had even *received a song*!

So, on that night, there at the window, I made a promise. From my time in the Listening Lodge, I made a pledge. Recalling the elder's words, I pledged to remain open to possibility. Remembering the buffalo's visit, I vowed to listen in a new way. Like some ordinary kitchen radio wishing to stretch out beyond its own pre-set limits, I committed myself to perceiving new frequencies on my radio dial. That night, in the middle of a snowstorm, with signals presumably crossed and transmissions temporarily down, I pledged to stay tuned.

❂ ❂ ❂

While the eighth story was written following my attendance at several weekend workshops, this story describes experiences that occurred during a number of discussions with Native American elders. The discussions centered around the Native American perspective of sound, as well as on the consideration of whether to build a sweat lodge. The eventual outcome of the sessions resulted in the decision not to build a sweat lodge. However, during the decision process, another kind of lodge came into being–a lodge built from listening, wherein by fully listening to one another a bridge was built, and openings for deeper kinds of hearing were created.

The Native American understanding of sound is rich and multi-faceted and there is no one "Native American" way of seeing or perceiving sound. Each of the tribal mythologies is profound and beautiful in its own right, though it is not possible to go into great individual detail here. "Native American" discussion of sound does, however, often focus on a worldwide web of sound. In Native American understanding, there is a web of relationship woven by the Creator into and through all of creation. This web is a sounding web, and each member of the living web sings its own song–the rocks, the trees, the birds, the insects, the people. Each has its own song, and each song supports the living of creation.

Creation itself is often described as having been sung into being. "The beginning was mist. The first Holy Ones talked and sang as always. They created light, night, and day. They sang into place the mountains, the rivers, plants, and animals. They sang us into life" (Luci Tapahonso, Navajo). "In the beginning, the earth was soft. There were no humans. There was no laughter. Born of Mother Earth and Father Sun are the humans. It is the Word of the Beginning" (Edmund J. Ladd). (Quotations from the exhibits at the Museum of Indian Arts and Culture, Santa Fe, New Mexico.) The land itself is song.

Likewise, in the Aboriginal beliefs of ancient Australia, creation is believed to have come about during the Dreamtime, when the great ancestors walked the land, depositing song trails (creation) into the terrain. "Indeed, each ancestral track is a sort of musical score that winds across the continent [. . .] an auditory route map through the country" (Abram, 166). Along with the auditory tracks, the ancestors are believed to have left behind trails of "spirit children" which enliven each fetus at the time of "quickening," thus giving each created individual a specific song derived directly from the land (*ibid.*).

In a Tiwa creation myth, it is written that:

> vibration [or word] decided that it would descend into form and become the living concreteness that we call the material form. It was in repeating its form over millions and millions and millions of moments and

light years that it was finally embodied, becoming what it had always loved to do the most and that was vibrating–vibrating as humans do and as energy does in the physical world or conceptual world, as the action states of various conscious awarenesses (Rael, 42).

Throughout the world's indigenous populations, sound is believed to be the driving impetus behind, as well as the stuff of, life itself.

Being created through sound, we also keep creation going through sound. We sustain creation through song, through ceremony. Each of us, made of vibration, is thought to be like a television set or a multiple-band radio, able to broadcast and receive many, many signals. If the electronics are sensitive enough, the equipment can pick up songs that are being broadcast–songs already sung from the past, songs to be sung in the future. Songs are all around us; it is only a matter of hearing them. However, most of us have our electronics set to receive a limited number of sounds. Like small kitchen radios, we pull in only a few fuzzy, local stations.

The ability to perceive past and future together may lie within the essence of Native languages themselves. Native languages have a different quality of relatedness about them than, for instance, English does, as they reside continually in the present moment of experience, partaking of the ability to change and to grow as new situations arise. Dr. A.C. Ross has noted, for example, that in the Dakota/Lakota languages, "time and space are inseparable. All temporal statements in the Dakota/Lakota languages are simultaneously spatial" (Ross, 56).

As with all gifts from the Creator, songs are medicine. As with other medicine (herbs or plants), songs are given for different purposes. Some songs are given for an entire tribe, some for curing specific diseases. Personal songs may be given to an individual for the purpose of healing one's self. Personal songs are sung by an individual throughout his or her life, as well as at the time of death, ensuring a balance between the singer and the Creator. Joseph Rael writes that "the chanter is the cosmic universe in miniature" (Rael, 119). Thus, in chanting, the singer unites earth and sky, created self with creating Creator.

> Medicine Singing goes back to the beginning of time with our people. They sang for survival; they sang when times were hard. [. . .] You have to ask yourself "Are you ready for it? Are you fit?" It's like this song is entwined around you and feeling you out to see whether you are a fit person for this song or not. There are many songs that will do that to you. There are many songs that will just come and bother you, until finally you sing it. Some of the songs come in dreams [Verbena Greene, Warm Springs] (Johnson, 152-3).

Medicine singing requires preparation and pure alignment of body, mind, and spirit. During ceremonial singing, spirits are called forth and asked to be present. When not prepared, when intoxicated, or when disrespectful of tradition, a singer may call forth negative spirits capable of doing harm during the ceremony. Likewise, when songs that are meant only to be sung as part of specific sacred ceremonials are sung out of context (for the entertainment of white audiences looking on, for instance), the consequences may be harmful to all involved. Far from the Western attitude in which music is seen as entertainment or as pleasurable experience, the Native outlook respects song as powerful medicine, capable of aligning the singer and listener with the primary forces of creation.

Medicine singing is capable of healing on many levels–healing the spirit, the body, or the mind. Medicine songs may speak to the individual; they may speak to the community. Medicine singing can bring about reconciliation and forgiveness. As the vibratory presence of the singer unites earth with sky, individual with Creator, forces that may have been alienated from each other are reunited. "Chanting is a way of making our past wrongs right, whether these wrongs are imagined, or real, personal or planetary" (Rael, 119).

The figure of the buffalo in the dream may refer to the depth to which we are called to listen, called to be in relationship with each other, as the buffalo is an animal that gives of itself completely. In Native American tradition, when the life of a buffalo is taken, all parts of the buffalo are used in support of human life. "Its flesh was food, its skin and hide provided material for clothing and for tipi homes. Sinews were used as thread, and the bones were made into needles, knives, and other implements" (Meadows, 168). The buffalo symbolizes the greatness of the Spirit that gives freely of its entirety for the life of all.

> In the Cree teachings, "The Listening" means more than anything else to us. The Cree Indian people learn how to listen to the environment, to the wind, to the rocks. We learn how to listen to everything. Some of the elders are saying that our young people need help to get back to "The Listening." [. . .] Our young people have forgotten. White people forgot a long time ago. They all need to come back and learn how to do this. There can be no real respect unless we learn how to listen to each other, not to hear what we want to hear, but to hear the truth [Vernon Harper, Cree] (Johnson, 130).

Native American prophecies tell of a time of purification when white people will return to seek the wisdom of indigenous peoples. It is said that during a time when the earth has become polluted and relationships have been rent asunder, the white people will come back to the Native peoples to learn again that sound is a sacred gift from the Creator, and that medicine singing is a path of healing for all creation.

CHAPTER 10
DAVID'S HARP

It had been a long time since I had seen my godchild, David. To be honest, it had been too long, because by now he was five and getting bigger every day.

When David was only a few hours old and still in the hospital, I helped give him his first bath while he exercised his full lung power for the first time. He vocalized further during his "first bath" at church, where I was honored to baptize him a few weeks later with all his family and church members smiling down upon him.

From the time he was born, David loved to sing. When he was sung to, you could see the sound vibrations wash over his tiny frame and he would respond, giving up any fussiness to the soothing, calming strains of lullaby. When David was a few months old, we would sing back and forth to each other. He had a high little voice, perfectly placed just above his two eyebrows. David lay, snuggled in his baby carrier on the kitchen counter, not much more than a foot long, and I would stand over him, looking down into his big blue eyes. I would sing a short, two- or three-note melody to him, searching, in a way, for some kind of communication from whomever it was who lived inside that tiny wriggling body, and he, in his little lilting voice, would sing the melody back to me. It seemed he was answering with affirming assurance that a small but powerful vibratory presence was now in the world. Sometimes we sang back and forth. Sometimes we sang together.

When David was about two and a half, I moved away. I missed him very much, and finally one day in July, I decided it was time to visit David in his new home.

When I arrived, David showed me all around his new house and his big backyard. He was especially excited to show me everything in his own room. One by one, he brought out all his favorite toys and put them on display–the action figures, the race cars, the bug house (with real live bugs in it!). Then there were the books. David was fond of books. We sorted through all the animal stories, the mice, the ducks, the dinosaurs. We looked at the word books; they showed lots of trucks and men with ladders.

Then, underneath a big pile of toys, we found a large, shiny picture book I had given to David once, which we had read together before. It was the story of another little boy named David, who lived long ago in Biblical times, who was a shepherd.

David, the shepherd, carefully watched his sheep every day. He was a good shepherd, and while he watched his sheep, he sang songs to them. While he was out in the hills and pastures, taking care of his sheep, he played on his harp and he made up songs for God. He sang about his deep love for God and about how God was his shepherd, and how God gave him everything that he needed. "The Lord is my shepherd," he sang. "I shall not want. He makes me lie down in green pastures and leads me beside still waters."

David, the shepherd, became very famous for his singing and his harp playing, and one day he was even asked to come and live with the king at his court. The king's name was Saul, and King Saul sometimes became very sick. David would go and play his harp for King Saul to try to make him feel better. David's harp playing was like the song of the angels. It was so soothing and so beautiful that it made King Saul feel better. Many years later, David, the shepherd, was to become the greatest king of all Israel.

"Aunt Cynthia, did you bring your harp with you?" David asked. He remembered the last time that I visited, I had brought a harp with me, and he had had fun plucking the strings and making them sing.

"I brought a little one with me this time. I'll show it to you later," I replied. "And," I said, in a voice filled with suspense, "I *also* want to tell you about a *very* special alphabet."

That afternoon, David and I went for a walk. There were woods behind his house, and it was a beautiful, warm day. The sun was shining, so we didn't need to wear our coats, but we took David's explorer belt so that he would be prepared. The belt had pockets attached to it to carry useful things like sunscreen, and it came with little bottles, in case we might find any good bugs.

We walked hand in hand from the bright sunlight into the darker shade of the woods, feeling the shift in air temperature as we crossed into the coolness below the canopy of overhanging trees.

"You know," I said, "a long, long time ago, when King David was a king, trees were thought to be holy. They were called 'sacred' because people thought that God lived especially in the trees."

"Maybe that's because their branches reach so high up in the sky, and they're so close to God," answered David.

"Yes, and their roots go so deep into the ground," I said.

"And," piped David, remembering his environment lessons from school, "they feed us by giving us oxygen so we can breathe!"

"That's prett-y sacred," I replied.

"Yeah, that's prett-y sacred," he said, mimicking exactly what I said in exactly the same tone of voice.

We continued to explore in the underbrush to look for special plants or strange-looking insects. We found a large mushroom shelf growing out of a tree trunk. It had a pretty pattern of brown and gray, but since it might be poisonous, we decided not to touch it and left it alone.

"Look, Aunt Cynthia," David cried, from somewhere behind me. I turned around to see David disappear over a fern-covered ridge into a large indented circular space that was lower than the rest of the ground.

"Gosh, look how big that circle is, David. Maybe Martians landed here."

"Yeah, or maybe pirates. Look what they left." David had found a big rock in the middle of the indentation, and he was pushing hard against it, trying to move it off its spot. "It won't budge! Maybe there's hidden treasure under here!"

"Or maybe a secret passageway that goes beneath the ground," I whispered.

"Oh," said David, in a low voice, as he stopped pushing on the rock, "maybe we don't want to go down there."

"No," I said, "but we could leave a message for whomever was here."

"Yeah, we could tell them that we've been here and that these are *my* woods now," said David, putting his hands triumphantly on his hips. Stomping his right leg and then his left, he began to strut around with a royal air, exclaiming: "I am David. I am king. And *I* am the *king* of *this* forest!" David's voice echoed from one tree to the next.

"You know another thing that people used to do a long, long time ago, aside from thinking that trees were sacred?" I asked. "They used the trees to talk to each other. The ancient Celtic people had a secret language and they would use one kind of leaf to stand for each letter of the alphabet. Then, they put the leaves onto a cord or onto a stick, and they would spell out messages to each other.

"But, for other people who didn't know the alphabet, or which leaf was which, it just looked like a bunch of leaves on a stick. But, you know, I think that whoever was here probably knew that tree alphabet. I bet we could plant your name in the middle of this circle and they would know who was here," I explained.

"But, what leaves do we use?" asked David.

"Well, I remember some of them," I replied. "For instance, we need an oak branch or some oak leaves for the two Ds in David. The oak tree was especially sacred. D was connected with the thunder and the lightning, and it stood for great power.

"The fir tree stood for the letter A, and the yew tree

stood for the letter I. Both the fir and the yew were connected with being able to live forever. That's why today we still use evergreen trees to decorate at Christmastime. The evergreen is a very old symbol of the ever-living Christ."

So we scouted around to find the leaves and branches that we needed. When we were finished, we had gathered two twigs of oak leaves, a small branch from the bottom of a fir tree, a medium-sized branch from a yew near the edge of the woods, and a heavy stalk of small, purplish flowers meant to stand for the V. With great seriousness, then, we pushed an oak twig into the ground, then the fir, the heather, and the yew, then, finally, the other twig of oak. With great ceremony, we planted each stalk into the ground behind the great rock, creating a row of leaves that spelled out the name: D-A-V-I-D.

"There!" exclaimed David, as he jumped up on top of the rock, throwing both arms up in victory. "Now they will know that I am David! And I am *king!*"

From somewhere far away, off beyond the edge of the woods, we could hear someone calling: "David . . . David." And as we listened closely, we realized it was David's mom. It was late and she was calling us home for dinner, so we followed our trail out of the woods and walked back to the house.

While we were eating, we heard all about how David was going to kindergarten this year. He had even gone to his new school and met his teacher. He liked her. Everyone told stories about what had happened since we had last been together, and then David asked his mom, "Can Aunt Cynthia tell me a story and tuck me into bed tonight?"

"That's a good idea," David's mom answered. "Maybe the two of you could even sing a song together—like you used to, when you were little."

We helped clear the dishes from the dinner table and then we helped David's mom get his younger brother, Patrick, ready for bed. Once Patrick had had his bath, we left his mom to tuck him into bed. Since David could stay up later than Patrick, we decided to get the little harp that I had brought with me.

David sat on the floor behind the small harp, just like King David had in the storybook. He plucked at the harp strings, each resounding with a bright, ringing tone. "You know," I said, "the Celtic people used to talk to each other on their harps. When each letter of the tree alphabet was assigned to a string, they could sit around their campfires and talk to each other in beautiful melodies late into the night."

"We could do that, Aunt Cynthia. We could play songs and we could stay up *all* night," David suggested eagerly.

"It was said that the harp could play three great tunes. One was for laughing." As I plucked out the melody to "Three Blind Mice," David laughed. "One was for when you were sad." As I played a sad tune, David pretended to cry. "And the last one was meant to put you to sleep!" With that, I started to pluck out the sweet, familiar notes of "Brahms' Lullaby."

"But, I'm not ready to go to bed!" David pleaded. Thinking he could keep me busy by telling stories, David added quickly: "But Aunt Cynthia, how did King *David* learn to play the harp?"

"I'm not sure. But he probably had someone in his family teach him when he was very young. Or, maybe, he just followed his heart, and listening carefully, he let the sounds he heard inside himself come out. Maybe the Lord taught him how to play," I answered. "No one knows. It was a very, very, long time ago."

"And how old is the harp, Aunt Cynthia?" struggled David in his attempt to stay up.

"The harp is very, very old, David," I answered, teasing him with a purposefully short answer.

"And, where did the harp come from?" David asked, hoping to distract me.

"Well, that would be a really, *really* long story. And, unless you get your pajamas on and climb into bed, I can't even begin to tell you about it." So David put on his pajamas and climbed into bed, settling in for the story of how the first harp came to be.

"No one really knows for sure," I began, "and there are lots of stories. Some say that the first harp came from India, some say from Egypt, and some say from the Middle East. But there's a very old story from the British Isles that reaches way back to the beginning of time, and it goes something like this:

"Once upon a time, in a far-off place, just after the world had been made and the mists were still swirling above the land in the cherub-pink and baby-blue colors of early creation, there lived a great Wizard.

"This Great Wizard was known far and wide and he was able to do fantastic, even miraculous things. When he spoke, he could make the wind stop blowing. When he sang, he could bring rain down from the skies. And, when he held out his hand, he could start a fire burning. Everyone knew and respected his great powers. People came from far and wide to listen to his advice. And they came when they were sick, because all he had to do was look at someone and he knew what would heal them.

"The Wizard had a very special well and folks came from miles around to be healed by its waters. The well was fed by springs that were said to be magical. The waters that bubbled up out of the ground at a very sacred place were said to come from the great mother of all creation. The sacred waters flowed into the magic springs, which fed the magic well, which the people drank to be free of what was ailing them."

"Why didn't he just play on the harp to make them feel better?" asked David, fluffing his pillows to be more comfortable.

"Well, we're getting to that," I answered.

"Now, all around this great magic well, there was a big group of hazel trees, called a hazel grove. The hazel tree was a very unusual tree, because, fed by the sacred waters, it held all the knowledge of the world, and the nuts from the hazel tree fell every day into the magic well.

"In this magic well, there lived some fish, and these fish were magic salmon. The salmon swam around in the water all day, watching the sacred hazel nuts from below. Then, just as a hazel nut would fall from the tree, a salmon would jump up out of the water, the sleek arch of its body gleaming in the sunlight, and it would neatly catch the hazel nut in its mouth and eat it before the nut could hit the water. Every time that a salmon ate a hazel nut, a red spot would then appear on its body, so you could tell just how many hazel nuts each salmon had eaten, and see just how much sacred knowledge that salmon had gained.

"Now, the Wizard wasn't satisfied with all the powers that he had. He thought that he needed more, and he wanted to have *all* the knowledge of *everything*. So, he had been watching this one great salmon very carefully for a long time. That salmon was quite old and it had the most red spots on its body, and it had obviously eaten many, many hazel nuts. The Wizard had been trying to catch this salmon for seven years and he knew that if he caught this salmon and ate it, he would then possess all the knowledge in the world.

"One day the Wizard lay in wait for the old salmon. The salmon moved slowly these days because he was full of knowledge and he was unhurried by things around him. As he swam in slow, serpentine fashion near the edge of the water, he couldn't see that the Wizard was waiting for him. When the old salmon lifted himself up out of the water in a great effort to catch the next hazel nut, he himself was caught by the greedy hands of the Wizard.

"The Wizard was ecstatic. At last he would have all the power of the world in his hands! He only had to cook and eat this one fish! He raced home and called for his young assistant, Finn. Finn had been studying with the Wizard for some time. Finn was very loyal, and he did exactly whatever his master, the Wizard, said to do.

"'Go and cook me this fish as fast as you can,' thundered the Wizard. 'But you, *boy*, are not to eat one bite of this fish! You must bring it directly to me!'

"So, Finn did exactly as his master had asked. He built a fire as quickly as he could, and he cooked the fish as quickly as he might. The fish was golden brown and piping hot as Finn flipped it from the frying pan onto a serving platter. The fish, though, was slippery, and Finn had to stop it from flip-flopping off the platter onto the ground. Finn put out his thumb to keep the fish on the platter, not realizing just how hot the fish really was. The minute Finn's thumb touched the fish, it seared his skin and burned him. Instinctively, before he even knew what he was doing, Finn stuck his thumb in his mouth to soothe the pain of the burn.

"By then, though, it was too late. The juices from the fish, which had oozed all around Finn's thumb, were now in Finn's mouth, making contact with his tongue. And all of the knowledge and all of the powers that the old salmon had held were transferred at that very moment into young Finn!

"The Wizard was furious and he flew into a rage, but there was nothing that he could do. It was an accident, of course. It wasn't meant to happen. But, now, in comparison with the ability of the young lad, the powers of the Wizard paled, and the Great Wizard knew that he would have to serve Finn as *his* master, all the rest of his life."

"But, what happened to the *fish*, Aunt Cynthia?" asked David, his head sinking deeper into the pillows, beginning to sound just a wee bit tired.

"Well, we're getting to that," I replied.

"Now the Wizard was no longer interested in the fish because Finn had already been given its knowledge. Having been given the soul, Finn ate the flesh, too. He enjoyed each and every bite, down to the very last morsel, leaving only a long, white skeleton of bone. When he had finished eating the fish, he tossed the bones over his shoulder and into the magical well.

"The moment that the bones came to rest in the well, the waters began to churn and swell. Bubbling into waves, great billows of sacred water surged up over the top of the well, each gurgling and laughing, and singing itself into rivers of deep, sacred water running every which way to the sea. It has been said that this was how all the sacred rivers of the British Isles came to be, and it has been said that the little skeleton of the fish was swept away by one of the most magical rivers of all, the River Boyne.

"The little fish bones were buoyed up on the tide of the sacred River Boyne as it gushed up from the well and out towards the sea. The bones rode atop the tide with the bubbling waters tickling over each rib. The waves babbled and murmured, rippling and gurgling, pulling songs from the bones that no human had heard. The bones sang out as they floated downstream, sending song after song to resound toward the skies. Then, slowly, the answers came back from the hillsides. Melodies floated from dale and from glen. The mountains rang out with songs high and glorious. The caverns bellowed the voice of the deep. The trees sang forth tunes that were many and varied, awakening the birds to enter the din. Soon, all of nature was sounding its glory, singing an orchestral song!

"The waves spewed up spray that mingled high into mist, while the clouds met the fog and reached down around land. You could not tell where spray-filled foam met cloud-swept sky. Then, just as the songs swirled up into climax, and just when the fish bones reached the mouth of the sea, they were whisked up high, out of the water, and set to spinning on the whistling wind. The bones became like a great white bird, fine and feathered, floating free. Then, out of the backbone came a golden frame, shaped with the grace of a swan, and each little riblet turned into a silver string, in a perfect scale to be played upon.

"That golden harp with the silver strings hung for a moment in the mists in the sky, and then it was lowered, as if by a loving hand, slowly down to the earth. There, on the bank of that sacred river, beside the well and next to the grove, appeared:

> "... *the swan-shaped harp*
> *with its tunes so eternal*
> *that its strains can be heard today.*
> *For that first harp*
> *was sent to bring love down to earth,*
> *and to take this world's pains away."*

"What did that harp sound like, Aunt Cynthia?" asked David, stretching out in a great, deep yawn, barely able to keep his eyes open.

"Well, we're getting to that," I replied again, as I reached out to pick up the little harp. "Maybe it sounded something like this..."

Strumming leisurely over the strings and humming lightly on the breath, I tried to imagine the strains of that ancient harp. Lost in the swirls of primordial mist, I heard the celestial melodies of the River Boyne. Playing them slowly, rhythmically, methodically, they merged with Logan lullabies rocking in the waves, and came to rest in a familiar tune:

"The water is wide, I cannot cross over, And neither have I wings to fly. Bring me a boat that can carry two, And both shall row my Love and I."

After a couple of verses, I paused to listen to the sounds around me. No bumps, no creaks, no voices could be heard from the entire house. Miraculously, for the first time all day, the house was quiet. Softly, I sang another verse, and then the last: "There is a ship that sails the sea. It's loaded deep as deep can be. But not as deep as the Love I'm in. I know not if I sink or swim."

Fading away, the final chords whispered, gradually melting one by one. Sitting peacefully in the silence, I listened expectantly in David's direction. Small, regular swells of breath rose and fell, but, lulled by the waves of resonant sound, not a single question came echoing back.

After all of the stories, following all of the song, David had fallen, at last, into the stillness that was to become tomorrow's dawn. Encircled now in the magic of harmony, David lay peacefully, fast asleep.

☀ ☀ ☀

This chapter touches on the subject of music and the Bible. In the Hebrew scriptures, music is often connected with ecstatic states of utterance and with prophets who receive the "Word of the Lord" in ecstatic or prophetic states induced with the help of music. Perhaps the most famous Bible story concerning music and healing is that of King Saul, who, plagued with what would seem to have been some form of mental disease or anguish, called in the young harpist, David, to soothe his ills with the gentle strains of his harp. While later becoming king himself, the stories of David's early life identify him as a sheep herder, a musician, and a composer. The Psalms themselves are attributed to David, the musician, and music is mentioned frequently in the context of worship within the Psalms.

The chapter then moves on to consider several aspects of how sound was used in the ancient Celtic cultures. Primary among these is the use of story itself (certainly not limited to the Celtic tradition), and a change in this story's tone is meant to signal to the reader a different, specifically "story-telling" voice, as if, for instance, one were reading to a child. The art of storytelling is used worldwide to entertain, to enchant, but most importantly, to transfer and to pass on cultural values and history. "The old traditional Celtic stories and songs recited by the bards were meant for both pleasure and instruction [. . .they] tell us about the Otherworld, about this world, about ourselves" (Matthews, 172). Within a culture's stories lies a cultural code, a telling of "who we are," "where we have come from," and "how we do things." In traditional societies, cultural identity and norms are handed down from one generation to the next through the telling of stories.

The very personal genre of storytelling, which often takes place in the context of one-on-one sharing from an elder to a younger family member, functions then on several levels. The use of sound, in this case, strengthens personal relationships while passing on inspiration, community identity, and group values. This very powerful use of sound as storyteller influenced the format of this book, it being finally written in the more-immediate style of personal story, as opposed to an academic, narrative format, with the hopes of achieving a closer connection with the readers. Storytelling is a sacred practice found in many traditional societies, and what better way could there be to transmit information about "sacred sound" than to use one of its own time-honored methods?

To ancient Celtic societies, the concept of "sacred" was often connected to and associated with Nature and Her forces. Trees were seen as sacred beings and as such were worshipped for their divine nature and their ability to guide one to the higher realms of the spirit. The "tree alphabet" mentioned here was connected to the ancient sacred language of Ogham (said to have magical power), and was actually used as a form of secret communication. Specific leaves, each standing for a specific letter, were strung together on poles or rods, spelling out messages for those to whom they were delivered.

The sacred tree of Celtic lore is connected worldwide with the universal Tree of Life. Many myths and legends portray this universal, cosmic tree as a musical tree. "Among all sonorous symbols, the tree is the most important because it is the ancestor of all cultic musical instruments. Whenever it is given a more precise description in the literature, it is portrayed as a hollow, singing tree that grows up out of a pond or a spring" (Schneider, "Acoustic," 73). The tree is often connected with the creative forces of the light of dawn, as well as with the morning dew: "Light spreads out from this tree, and, through the process of synaesthesia, it has come to be known as the 'singing tree' of legend and folklore" (Cirlot, 81).

In many Eastern traditions, the Tree of Life is closely connected to both traditions of the *Logos* and the Goddess:

> A Goddess of the Tree was worshipped as early as 2000 B.C. in the Indus Valley, where the tree was called the source of all *mantras*–hence, of creation. The tree gave the gift of language to humanity, having a different letter or *mantra* written on each of its leaves (Walker, 473).

In southeastern Asia, a Tree of Life is described in which a "destiny word" appears on every leaf:

> Such words became personal *mantras* for all souls of the dead, each of whom must obtain one of the tree's leaves before rebirth to a new life on earth. It was claimed that the word on the leaf determines the character of each new life (*ibid.*, 474).

The connection between the Eastern singing tree with its mantric leaves, chosen by incarnating souls, is interesting in that it parallels a similar sonic process of incarnation brought about through the Western harmony of the spheres, as described earlier in the Greek story of "The Myth of Er."

The re-telling of the story of Finn MacCumhail, in a kind of fanciful, free-form style, incorporates a number of magical symbols found woven throughout Celtic lore—the hazel grove, the hazel nut, the sacred well, the salmon, each a highly revered bearer of sacred and mystical knowledge in its own right—along with one greedy old Wizard and one young, valiant apprentice, the latter of whom (of course) wins out. Throughout this chapter, I am indebted to conversations concerning Celtic matters with Richard Gary. As part of the story, the bones from the wisdom-granting fish transform themselves via a journey through sacred water into the strings and the body of the mystical harp, whose magical strains are said to hold the power to heal. For this telling, I am likewise indebted to his wonderful poem, "The Harp of the Boyne."

The healing harp is prevalent throughout Celtic myth as an instrument of higher consciousness and power, its enchantments having come directly from the chief of the Celtic gods, Dagda. Dagda, a master musician, is said to have initiated the "three noble strains" played by the harp: "The strain of lament or sorrow caused the release of tears, and the laughter strain raised the spirits of the listener. The sleep strain brought rest to troubled souls and plunged them into profound slumber" (Williams, 12). This chapter, thus, ends with the "noble strain" of sleep, closing as the harp spins forth its line of hypnotic intent into a "lull-a-bye," and the young apprentice of the present story, enchanted himself by music's power, is lulled a-bye.

> Those who were sung to as children remember loving the intimacy which lullabyes bring. They felt special because they knew that 'the songs were just for me.'[...] Even off-key singers are appreciated by babies and children because love and warmth are felt through the song. It doesn't matter if the notes aren't quite right. It really doesn't matter what you sing. What matters is the love (Hale, 60).

CHAPTER 11
REMEMBERING THE NAMES OF GOD

The setting was a grand old mansion with rolling lawns, several sets of out-buildings, and two-hundred-year-old trees that sheltered a number of walking paths that wound their way around the property. The mansion itself had seen better days, but was now being lovingly, though sparsely, maintained as a spiritual retreat center. A shade of its former self, the mansion was still gracious and welcoming, with its large bedrooms converted into dormitories and its grand parlors and sitting rooms now serving as comfortable meeting spaces for the spiritual pilgrims who assembled there.

The first *zikr* from the Sufi tradition, to which we were introduced, was a walking *zikr*. In Sufism, a *zikr* is a devotional practice that involves repeating a name of God continuously, with the intention of bringing one's full attention to the remembrance that God's presence is found deep within one's being at every instant. To remember continually is to be touched continually by divine forces that enliven the heart, and remembering this brings bliss and the blessings of divine communion. To "pray without ceasing" the Name of God is considered, in many spiritual traditions, to be the door that opens the inner sanctuary of the chanter to the holy of holies.

The group was instructed to complete a leisurely circuit, beginning at the big house and following the walking paths to end up back where we began. While walking, we were to repeat silently and continuously one of the divine Names of God. While the phrase *"Allah hu"* was suggested from the Islamic Sufi tradition, we were given the choice of choosing a name with which we would be comfortable, from a tradition of our choosing, and were asked to be aware in each moment how that name brought us into closer communication with the Divine Presence.

Having been given the latitude of making a choice, I descended the cracked cement steps that led from the porch to the large, circular driveway with my footsteps crunching rhythmically across the gravel and my mind recounting in time the almost infinite number of names from which I might choose. I thought of the Islamic traditions that used the Ninety-nine Beautiful Names of God, allowing meditation on such attributes as the Merciful, the All-peaceable, the Holy One, the Wise, the Benevolent, the All-hearing, the All-seeing, the Just, the Very-strong, the All-forgiving, the Friend. I thought of the Hindu traditions in which a thousand names are used to honor the Mother or to worship Vishnu, with such contemplative descriptors as "all-pervading," "supreme light of consciousness," "sanctifier of what is most sacred," "who controls everything," "who removes all sorrows," and "giver of relief and shelter." I thought of the prophet David, who proclaimed through the Psalms: "Yahweh, our Lord, how great is Your Name throughout the earth!" and the Sikh founder, Guru Nanak, who wrote, "Happy are the saints who, listening to His Name, destroy suffering and sin."

I chose to begin with the ancient syllable "aum." As a traditional introduction to any *mantra*, whose purpose is to open sacred space, it was an appropriate place to start. Chanted for centuries, so rich in meaning, it encompassed all the sounds and all the Names in what it named. It crossed all worlds and mediated all levels of being. Reaching from the grossest of the manifest to the subtlest of the Unmanifest, it was certainly one sound that continually called to mind the sacredness of all being and the closeness of the Divine.

Then, reaching a point on the path where it shifted off towards the right to run parallel with the property line, I shifted internally to chanting another favorite *mantra*, the Name of God as one of the incarnations of Vishnu (the Sustainer), that of Rama. I repeated: "*Rama, Rama, Rama, Rama–Rama, Rama, Raa-aam. Rama, Rama, Rama, Rama–Rama, Rama, Raa-aam.*" This was another ancient and sacred Name; this was the name that Mahatma Gandhi had used to sustain himself. As he sat spinning cloth in his cottage, he had silently repeated "*Rama, Rama, Rama, Rama–Rama, Rama, Raa-aam.*" As he traveled and taught throughout the world, he had silently repeated "*Rama, Rama, Rama, Rama–Rama, Rama, Raa-aam.*" As he led his country out of colonial bondage into what he hoped would be a peaceful bid for freedom, he had ceaselessly prayed the name of God: "*Rama, Rama, Rama, Rama–Rama, Rama, Raa-aam.*" His final words, uttered as he lay dying, downed by the bullet shot from an assassin's gun, were "*Hey, Raam.*"

The beauty of the late spring afternoon spoke to me from everywhere along the path. The birds called out to one another, singing their celestial songs. The wind sang lightly through the branches of the trees; the warmth of the sun vibrated down into the spongy, new green grass, as my feet left regularly repeated imprints upon it: right, left, right, left–*Rama, Rama, Rama, Rama*–right, left, right, left–*Rama, Rama, Raa-am.*

The Name of Rama was also known as the *Taraka mantra*, a *mantra* that carries one across the ocean of transmigration, capable of freeing the soul from its endless returnings. So many sacred traditions echo this belief in the connection between the Name of God and a sure passage at the time of death. I thought of "om," long referred to as a ferry-boat used to cross the ocean of life and reach the peace of the Eternal shore. I thought of the chant of the Western rosary, "Holy Mary, Mother of God, pray for us sinners, now and at the hour of our death," prayed so often at the bedside of someone sick or dying–another witness to the worldwide belief that if one died with the Name of God on his or her lips that he or she would be received immediately into God's loving arms. Richard Rolle, the Western mystic, wrote in the *Fire of Love:* "He who all his life has based his meditation on that lovely Name will die surrounded by marvelous melody."

Of course, the "lovely Name" that gives rise to that "marvelous melody" in the Christian West is the Holy Name of Jesus, for in Western theology Jesus Christ is the Word, and Jesus is the Name-above-all-names. In the Hesychastic traditions of the Eastern Orthodox Church, the Jesus Prayer has long been prayed for its beneficent effects. Perhaps, when prayed in Latin or Greek, or even in Old Church Slavonic, the sounds of that prayer were comforting and soothing, but, personally, I had always found the English sounds and tones in the phrase "Lord Jesus Christ, have mercy on me, a sinner" slightly awkward and halting, as if they were all colliding into or stumbling over one another.

So, as I turned to follow the curves of the final leg of the path, I shifted into chanting a name that had a more flowing quality in its Latin intonations, one I had always found to be quite soothing: *"Jesu Juva." Jesu Juva* was the phrase found on many of the pages of J.S. Bach's musical manuscripts. *Jesu Juva.* With the soft sounds of the Latin letter J sounding like our letter Y, or "ya," and meaning "Jesus help," this must have been Bach's continual prayer: *"Jesu Juva, Jesu Juva, Jesu Juva."* Sometimes written only in initials (J.J.), this must have been his personal petition, a flowing repetition of the Lord's name. As the intoxicating melodies of inspiration flowed from the Divine Mind into that very-human container, so fast, in fact, that he could barely get one phrase written down before the next harmonic inspiration had begun, I imagine this *mantra* must have been his centering calm.

I had relaxed into the soothing, melting sounds of *"Jesu Juva, Jesu Juva"* as I completed the final leg of the *zikr.* My gait had synchronized with the elongated sounds, my pulse beat in time with my respirations as they rose and fell with each repetition of the silent chant. *"Jesu Juva. Jesu Juva."* My mind was quieted now, like a lake, once worried by the wind's sporadic ruffling. *"Jesu Juva. Jesu Juva."* Like a lake whose surface has become so completely calm you can see clearly to the bottom, so had my mind become one with the Name. *"Jesu Juva. Jesu Juva."* Seeing clearly through the water's surface, I felt the presence of that Loving Light enshrined deeply within.

As I left the natural world of the walking path, I recalled the words of a hymn paraphrasing the 23rd Psalm: "The King of Love my shepherd is, whose goodness faileth never; I nothing lack if I am his, and he is mine for ever." The image of the stilled lake mixed with the words of the next verse: "Where streams of living water flow, my ransomed soul he leadeth." And, leaving behind the lovely, new greens of the springtime fields, I heard "and where the verdant pastures grow, with food celestial feedeth."

As the images from the present moment merged with those of the song, it was as if I were remembering that there had always been a Holy Presence with me whose steps were synchronized with mine, guiding my footsteps along the walking paths, soothing me with the stilled sounds of living water and celestial song, bringing me eventually back home. Just as the melody had arisen so easily and so freely from somewhere within, so the awareness of that Loving Presence had arisen quite naturally, simply by bringing my full attention to the Name. Crossing rhythmically back over the gravel of the driveway, I ended my journey, humming the tune from the closing verse, "Good Shepherd, may I sing thy praise," just in time to ascend the stairs and to set foot on the front porch of the old mansion, "within thy house forever."

The others returned also, one by one, to an area on the front porch, each having completed his or her own journey silently chanting a Name of their own choosing, and then, for the second *zikr*, we moved together indoors into one of the spacious parlor rooms. There, where all the furniture had been removed, the high ceilings, wide windows, and smooth floor-boards offered a perfect place for the twenty or so of us to join hands in a circle and begin to move slowly and methodically to the left.

The mesmerizing rhythms of the frame drum filled the space of the room, as the leader chanted the words "*La illaha*" (there is no god), "*Il Allah*" (but God). "*La illaha–Il Allah.*" Synchronizing our breathing, we exhaled to the words "*La illaha*"; we inhaled to the words "*Il Allah.*" "*La illaha–Il Allah.*" Matching our steps to the tempo of the drum, we moved together, rocking gently forward, rocking gently back. "*La illaha–Il Allah.*" With each exhalation, we breathed out tension, distraction, any disturbing thought. "*La illaha–Il Allah.*" With each inhalation we breathed in an openness to the Divine Presence, we breathed in the air of peace. "*La illaha–Il Allah.*" Exhaling, we emptied ourselves of selfishness; we cleansed ourselves of toxins. "*La illaha–Il Allah.*" Inhaling, we breathed in the clarity, the purity, the Presence of the One. "*La illaha–Il Allah. La illaha–Il Allah. La illaha–Il Allah.*"

Looking down, I saw first my left foot and then my right. The left foot moved, the right foot followed. The left foot took the weight, crossed over by the right. "*La illaha–Il Allah.*" First the left, and then the right. The group rocked rhythmically back and forth, continually moving towards the left. "*La illaha–Il Allah.*" Our heads began to move together, first to the left, and then to the right. "*La illaha–Il Allah.*" Holding on to each other's hands, poised on the verge of imbalance, dependent on each other for coherence, we continued to circle round and round. "*La illaha–Il Allah. La illaha–Il Allah.*"

At some point, centripetal force took over. We were no longer twenty people holding hands and moving in a circle; we were a finely oiled, well-tuned wheel, spinning precisely in time to the chant. Had we even *wanted* to stop, we were lost in the movements of the wheel's turnings, as our lungs refined the respirations and our muscles strained to keep the time. With every cell and neural pathway, our bodies themselves met the rhythm and sang out the phrase: "*La illaha–Il Allah. La illaha–Il Allah.*"

The rhythms of the drumming picked up speed. This was no mental exercise, but a corporeal declaration. "*La illaha–Il Allah.*" This was no philosophical argument, but a physical, heartfelt exclamation. "*La illaha–Il Allah.*" This was no quiet retreat to a safely hidden garden pew, but an open expression of a joyful noise. "*La illaha–Il Allah.*" There was no denial of the body here, but an unqualified celebration of the rhythms of life. "*La illaha–Il Allah.*"

In time, the pace slowed down. The movements diminished and we came gradually to a halt. Dropping hands, we bowed to each other, and began to move into the sitting room next door. As we found our seats, we were still tingling. Our muscles and sinews, our skin and our nerves still resounded with jingling and jangling, as our bodies reverberated the echoes of the powerful chant. Resting now, sitting perfectly still, my body continued to trace inwardly the movements of the words, while my mind sang out the tune: "There is no god but God."

"*La illaha–Il Allah*," my spirit answered back.

Following a brief reflection on how people found that the physical activity of the two *zikrs* had brought devotion directly into their bodies in such a way that an act of worship had actually taken place within the body itself, we moved on to a third *zikr*, which did not involve outward movement, but a concentration on inner movements–movements of the soul. We began an inner practice intended to further remove barriers to the divine presence, intended to "polish the mirror of the heart."

For this exercise, which consisted of two parts, we

closed our eyes. First, we were to imagine a bright, shining light in the area of the Third Eye while inhaling and chanting the Name "*Allah.*" Then, we were to envision sending that light down into a small area within the heart, while exhaling and chanting the syllable "*Hu.*" Coordinating the breath with the chant, we were to continue this visualization unvaryingly: "*Allah–Hu. Allah–Hu.*"

Allah, we were told, was a certain indefinable awareness of divinity, the sensation of being wrapped in the compassion of a loving Friend. "*Allah–Hu.*" Hu was a special awareness of that Presence. "*Allah–Hu.*" With each repetition we were polishing the reflection of a compassionate Friend deep within ourselves. "*Allah–Hu. Allah–Hu.*"

In the abrasion of each inhalation, we polished our spirits, breathing in love, light, and compassion. "*Allah–Hu.*" Like rust being removed from iron, we exhaled malice and spite, prejudice and hatred. "*Allah–Hu.*" Swirling currents bathed our minds in clear light, which was, in turn, sent down to the heart. "*Allah–Hu.*" Revenge and envy disappeared like germs dried up in the heat of the sun. "*Allah–Hu.*"

Our individual breathing patterns joined into one, entraining with each other in time to the rhythm of the chant. All our voices mingled together; the individual qualities of resonance became one radiating column of sound as we poured light from the Third Eye down into the heart. "*Allah–Hu. Allah–Hu.*" The sounds and the rhythms merged into one long gesture. "*Allah–Hu. Allah–Hu.*"

A- was the first sound to enter the body, vibrating and moving down the back of the spine. *-llah* wrapped around the base of the root chakra, joining itself deeply to the body. Releasing then, the *Ha-* rose to re-enter the throat, while *oo* left the lips, after resonating the roof of the mouth and carrying vibrations to the brain. "*Allah–Hu. Allah–Hu.*" We were creating a circle of sound. A circular polishing was being completed, while light from above was pumped down below, the sounds themselves making repeated turnings. A circular pumping was becoming established, formed by the group in its repetition of sound. "*Allah–Hu. Allah–Hu.*"

With inner eyes, I was watching the light flow. Cascades of clear and luminous light fell from the heights, collecting into pools like water below. The liquid pools then circulated into tunnels, following the contours of walls from one channel into the next. The rushing of liquid was rounded and regular, shaped by the walls in the tunnels of sound. Suddenly, the translucence gave way to color and I was flooded with the sensation of bright, red liquid spewed out from something, being pushed regularly forward in spurts. Gushing and rushing, it passed in eruptions from one contained chamber to the other, surging and receding in a rhythmic flow. "*Allah–Hu.*" The tissues contracted, walls tightly constricted; then muscles relaxed to allow the flow. "*Allah–Hu. Allah–Hu.*"

Red liquid soaked into the tissues, cleansing the muscles in rhythmic song. The sound of the liquid pumped through the chambers, echoing the passage of the return each time. "*Allah–Hu.*" Kinesthetically, I could feel the mechanical beatings of a huge heart. This was the heart of the Supreme Being. *Allah.* This was the heart of All People. *Hu.* These muscles recalled the actions of every creature who had come into being; these walls pulsed with the memory of everyone who had ever been born: "*Allah–Hu.*" This blood ran, laughing, through eternal chambers, while it carried memories, crying out in pain. The darkness of the pumping recalled all the sadness, while the light in the liquid brought in joy reclaimed. "*Allah–Hu. Allah–Hu.*"

In the midst of the chant, our group had united. We had, in effect, all become one. Dropping the sounds that had served to separate us, we had moved to a place of sensing beyond the Names. There we moved beyond "Allah," beyond "Rama," beyond "Jesu." Releasing words like "Muslim" and "Hindu," we dropped mental constructs so as to sense how false things "outside" were "inside" true. There we remembered a sacred origin, where all names rise to praise the One.

Allah–Hu. Al-le-lu-i-a. Allah–Hu. All-e-lu.

Physically feeling the heart of the Center, we remembered to recall what we had forgotten to name. This heart is the True Friend: *Allah.* This heart is the Presence: *Hu.* This heart is the cause of our being: *Allah.* This heart brings remembrance: *Hu.* This is the Heart deep within my own heart: *Allah.* This is the Heart deep within you: *Hu.* These are the sounds that polish the mirror: *Allah.* This is the moment: *Hu.*

Allah–Hu. Al-le-lu-i-a. Allah–Hu. All-e-lu.

ALLAH–HU!

❂ ❂ ❂

This story investigates practices that are associated with the Sufi spiritual path, although they are found in other spiritual teachings as well. As with the chapter on shamanism, this story gives the reader only a taste of experience gathered through attendance at a spiritual retreat. While deep understanding of any tradition can come only with sustained practice over a long period of time, the exposure to this spiritual tradition, which (like shamanism) uses the medium of sound as an opening for spiritual awareness, was quite profound, even if encountered for only a short time.

The Sufi orders, very devout sects found within Islam, are devoted to direct and ecstatic experience of the divine. And here, the "[e]cstasy is not a feeling to be pursued; rather it is the freedom from all self-seeking pursuits," an "ecstasy of selflessness" (Helminski, "Ecstasy," 2–3). The word *sufi* is connected to the white wool garments often worn (Kramer, 163), as well as to the word *sophia*, wisdom of the inner worlds (Khan, *Music of Life*, 235). "The Sufi and [Dervish] conception of music as an aid or approach to religion is of the highest importance as [...] Ibn Zaila (d. 1148) says: 'Sound produces an influence in the soul in two ways: one on account of its musical structure (i.e., its physical structure), and the other because of its similarity to the soul (i.e., its spiritual structure)" (Schneider, *New*, 440). Many Arabic mystics speak of music as the primary route of ecstasy that lifts the veil, leading the worshipper into the heart of Allah.

Hazrat Inayat Khan, the great Sufi mystic and musician of the soul, writes beautifully of a cosmology grounded in music: "My sole origin is sound. [. . .] The Sufi names music *ghiza-e ruh*, the food of the soul, and he uses it as a source of spiritual perfection. For music fans the fire of the heart, and the flame arising from it illumines the soul. [. . .] Music is the picture of our Beloved and our Beloved is that which is our source and our goal" (Khan, *Music of Life*, 5 & 58). Elsewhere, he states: "The mystics of all ages have loved music most. In almost all the circles of the inner cult, in whatever part of the world, music seems to be the center of the cult or ceremony" (ibid., 121). This is why the Sufis meditate with music: "The effect that they experience is the unfolding of the soul, the opening of the intuitive faculties; and their heart, so to speak, opens to all the beauty which is within and without, uplifting them, and at the same time bringing them that perfection for which every soul yearns" (ibid.).

The Sufi spiritual practice of *zikr* (or *dhikr*) involves the meditative repetition of the divine names. This term *dhikr* means "recollection," and it is through constant recitation of the names of Allah that the heart is filled with nothing but the all-encompassing presence of divine love. The practice can be done sitting; it can be performed alone, or carried out with a group of people in a circle, or in other geoetric forms. The names can be repeated while whirling, as the dervishes practice, or the name(s) can be "recollected" during a walking meditation. "[T]he action or rhythm of walking [is] a technique of dissolving the attachments of the world and allowing men to lose themselves in God. The aim of the dervish [is] to become a 'dead man walking,' one whose body stays alive on the earth yet whose soul is already in Heaven, [thus becoming] the Way not the wayfarer, i.e., a place over which something is passing, not a traveler following his own free will" (Chatwin, 179).

Centered in the heart, the Sufi tradition emphasizes the physicality of these practices, the results of which are to be felt and experienced in the body. The changes that occur take place on a physical level, as well as on the spiritual plane:

> The whole mechanism, the muscles, the blood, the circulation, the nerves, are all moved by the power of vibration. As there is resonance for every sound, so the human body is a living resonator for sound [. . .] there is no greater and more living resonator of sound than the human body. Sound has an effect on each atom of the body, for each atom resounds; on all glands, on the circulation of the blood and on pulsation, sound has an effect (Khan, *Music*, 41).

This remembering of the Name, whether accompanied by physical movement or not, is prevalent in many spiritual traditions, from the reciting of the "Hail Mary" and the Jesus Prayer in the Christian tradition, to the *Taraka mantra* in the Hindu tradition. From the Christian tradition come the words of St. John Chrysostom: "The names of Jesus Christ is terrible for demons, passions of the the soul, and diseases. Let us adorn and protect ourselves with it"; and from the Hindu scriptures, the *Shrimad Bhagavatam*, come the words: "Unknowingly or knowingly the chanting of the

Supreme, praiseworthy Name burns away man's sin, even as fire reduces to ashes." The practices are similar, though called by different names. Christians may simply use the word prayer, or say, in the words of St. Paul, "to pray without ceasing." Hindus refer to *mantra yoga* or *mantra shaastra*, while Sufis use the term *wazifa* or *zikr*. In the philosophy of the Path of the Masters, wherein the soul awakens to the divine through sound, it is called *simran*, derived from the Sanskrit root *smr*, to remember, to recall, to protect. "By constant *Simran* one awakens superconciousness and attains the state of everlasting tranquility and peace" (Singh, S. I, 66). Or, "[a]s Zeb-un-Nissa says, 'Say continually that sacred name which will make thee sacred.' [. . .] It is the power of the word which works upon each atom of the body, making it sonorous, making it a medium of communication between the external life and the inner life" (ibid., 70).

Biblical instances have been mentioned above which refer to ancient Hebrew practices that connect the power of music and sound to meditation and superconsciousness. "The Baal Shem tov, the founder of Chasidism, taught 'Through music you can reach joy and union with the Infinite One,'" and "[t]he great master Rebbe Nachman of Bratslav used to teach that 'through music you can come to the level of prophecy. For the essence of devekut (rapturous attachment) with God is through melody" (Davis, 21 & 140). Likewise, in the mystical practices of the *Kabbalah*, it is through contemplation of the Names of God that one works through the spiritual aspects of the Tree of Life, reaching the higher realms, and, ultimately, the goal of *Ayn Soph*.

"It is only *Rama Naama* which brings peace and enables us to swim across the world-ocean," (Singh, J., 173) writes Guru Nanak. For the fifteenth-century founder of Sikhism, the repetition of the Name is the highest teacher, for the Name actually *is God*, and thus *Nam* itself is the highest mystical principle, as well as the ultimate *sadguru* (true teacher).

He speaks of this most sacred truth in his *Siddh-Gosht*:
> *As the lotus flower*
> *Does not drown in the pool,*
> *As the duck swims,*
> *So with the mind, intent*
> *Upon the word of the Guru,*
> *One can safely cross*
> *The great sea of life,*
> *Repeating the Holy Name.*

CHAPTER 12

ANCIENT AND SACRED –
Nothing but the Sound

New York City was overwhelming. People packed on top of people jostled and jammed their way past each other. The blinking of blinding signs flashed sales of the latest fads. Acrid smells rose from filth left to rot in the open air. Nerves strained on the edge of snapping, frazzled from years spent in the electric wear and tear. It was inconceivable, but the city had somehow managed to intensify since I had been here last, bombarding my senses with the very best and the very worst of everything.

I noticed, too, that I was nervous. Having to watch constantly, looking to the left, looking to the right, questioning who was behind me, what someone might do. My fingers drummed nervously while riding the train into the city; my foot tapped impatiently while waiting for the lights. The constant over-stimulation had a way of sweeping people up into frenzy. It had a way of magnifying an inner restlessness, which, of late, had held me captive, pacing back and forth, prowling from within.

It was as if I were trying to complete something, to bring an end to things begun. Having vowed to return to the music, I had been on an unnamed quest I couldn't describe, a search for sounds I couldn't hear. Touched by voices in a singing web, I had become aware of something sounding so deeply that it called continually to me, like a magnet pulling its object through invisible fields.

I had come here on a kind of harmonic hunch and I was looking forward to what might unfold. It's funny how harmonic intuition doesn't always make logical sense, I thought, as I waited to cross the street to attend a workshop on something about Sanskrit, a presentation on ancient sound. With horns blaring forth the deafening forces of modern-metropolitan din, this didn't seem, at first glance, to be the most likely place to break through any barriers into the ancient realms of sound.

Tired from travel, I crossed the street, tracing my way from one building to the next, finally reaching the address I had been given. My soul responded in welcomed relief as I turned a corner, swung open the great glass door to leave behind the glitz and the clamor, and climbed the four sets of quiet inner stairs, and entered gradually into the calm of a large studio with many rooms equipped and arranged for teaching yoga. In here, there was a different sense of space, and I could feel the change.

The voice I heard welcoming me was one I identified immediately, but I couldn't have told you why. The vocal tones were warm and recognizable, reminiscent of a distant but acutely familiar past. I returned the welcome with equal warmth and enthusiasm, as I became aware of the sensation of being carefully wrapped by a blanket of unspoken knowing, along with the distinct feeling that, after many years of having been away, I was finally coming home. A melody floated lightly into my awareness. "And the voice I hear, falling on my ear . . ." wafted gracefully through my mind, quelling my questions. So I am supposed to be here. My eyes slowly surveyed the room and I chose a place to sit where I felt comfortable as others entered the room, joining the class, sitting primarily on the floor, forming a semi-circle several rows deep around a centrally placed easel.

As the lessons in the Sanskrit language began, we were introduced at the same time to some basic but challenging lessons in educational de-programming. We were not, for instance, allowed to take written notes. "Not allowed to take notes?" I asked. "But how will I remember anything?"

"You don't have to worry about remembering anything," came the answer.

"But I won't be able to memorize what you're saying," I protested.

"You don't need to memorize for this class. Just show up."

No notes? No memory? What was this? "But how will I know what I'm learning? Or what I'm able to take away?" I asked.

"This isn't about piling things up, collecting them, or taking them away," came the amused answer. "You will take plenty with you. You simply have to trust me in this."

Trust you, I thought, I don't even *know* you! The melody from the hymn "Awake, My Soul, Stretch Every Nerve" danced liltingly across the top of my mind, ending with the downward-drifting phrase "and press with vigor on." It *would* be a stretch, I thought. However, I *was here*, and there did seem to be some reason for it. I made a conscious choice to stay.

The real test came when we were asked to go around the room, one at a time, chanting a series of nonsense syllables out loud. The tension in the room grew like a great gray cloud; everyone's nerves stood on tiptoe as they strained to repeat slippery sounds that made no sense and seemed to defy memory. I found that I was all right with the first two sets of sounds; I could hold a certain amount of information in my mind. But with the addition of a third and then a fourth set of incomprehensible material, I had trouble holding on to everything that I'd managed to remember from before. The minute I'd concentrate on the fourth line, I'd lose the first. If I focused on the first line, I'd forget the fourth. It was frightening to be aware of all the sounds that I couldn't possibly balance, corral, or contain, but which I was expected to repeat back in front of the whole class!

"You don't have to get it right," was the instruction. "If you forget some of the sounds, feel free to make up something."

I don't have to get it right? I thought. That'll be the day! Years of music lessons flashed through my mind in little vignettes: fingers placed perfectly in just the right position, the beautifully executed vowel, the nerves of public performance. Time! I thought frantically, I just need more time! I need to hear the sounds just a few more times! Quickly I reviewed all the sounds, using every rehearsal technique I could summon.

"What's going on besides just listening?" came the teacher's question. "What's going on other than just listening to the sound?"

"Well, there's a *lot* going on in here other than just listening," I replied. "I'm trying to figure out how to hold on to all this information in just the right sequence in a certain part of my brain for exactly the right amount of time so that I can instantly recall on demand all of the details in precisely the right order."

"But you don't have to get it right. Make up something if you need to. Try and listen just to the chanting as it goes around the room. Focus only on the one sound that is being made, and when it comes to your turn, focus on that sound, like any other sound, in just the same way. But remember, focus just on the sound and nothing but the sound."

"Just the sound," I repeated, "and nothing but the sound. Just the sound and nothing but the sound. Just the sound and nothing but the sound." Eventually, even the words "nothing but the sound" faded from my awareness as I shifted my attention, dropping my anxiety and focusing only on whomever was chanting at the time. In the process, I became aware of a subtle but profound shift in the room. The great gray cloud that had formed earlier was beginning to subside; people were becoming more relaxed and at ease. We were able to give full attention to each person as he or she chanted and were not worried about ourselves.

As I dropped my own concerns, my mind became quieter and I could listen from a place of inner contentment. From that place of silence, each voice that I heard resonated with a vitality that was truly captivating. Like a necklace of sonic jewels, each voice opened up, one after the other, each sparkling with its own unique qualities and intriguing color. And, as each gem revealed itself, it was like listening to a beautiful and dazzling necklace of sound. When it came time for me to chant my own link in the chain, no longer nervous, I was amazed at how much I could chant back when not focused on thinking about it.

Listening deeply, it was as if I were inside of whomever was chanting, sounding like them, sounding with them. We had moved from being a number of disconnected, self-conscious individuals into being one entity that was resounding beautifully, functioning as a whole. Connected in some deep and quiet space, we continued chanting like a radiant sonic necklace, laughing when someone made up something ridiculous, breathing relief when someone finished a rough patch, relaxing as the group grew more deeply connected and more comfortable with each other. We closed that first session with a Sanskrit verse, and the strange, incomprehensible words were ended with the instructor's own chanting of the syllable "om."

As I listened with eyes closed, relaxed yet alert, the word "om" appeared before me in the darkness. Shaped something like an atom, something like the sun, with trajectories of circling color and swirling light, this living, breathing being unfolded before my inner eyes. Formed from invisible beams of sound shot through open space, the vibrations from the word translated themselves into color and form. It radiated vitality. I could have held it in my hands. It was as if this "om" were alive! Mesmerized, I sat lost in the depths of inner space and transfixed by the colorful echoes being spun forth from that living light, long after the physical vibrations had faded and the class had been dismissed for the night.

When we returned the next day to resume our study of the Sanskrit alphabet, it did seem to me that I was hearing a little differently, maybe a little more clearly or in a more focused way. Sounds had become multi-dimensional, composed of several layers. I could pay attention to the outer rim of a word, or I could sink deeply through its inner workings into the source of its energies. We were told that we would be working with an ancient yogic process, the primary formula of which sounded like *"yogash-chitta-vritti-nirodhah."* Even as we heard the formula for the first time, even as we focused on the meaning of the words, there seemed to be a force at work within the phrase itself. While we learned that the root word in *yoga* could mean a kind of union with the divine, and while we engaged in the didactic exercise of translating the words into English (*"yoga* is the selective elimination of mental activity in the field of consciousness") there seemed to be a larger force at work, moving below the gross level of the sounds, which was effecting a change, bringing a clarity to our inner hearing and a lucidity to our inner sight.

As we began to learn the sounds of the alphabet, we were introduced to the vowels of the language. They were like newfound, long-lost friends–energy fields with whom we were getting re-acquainted. We asked questions; we played. The primary sound of the alphabet was said to be the sound uttered first at creation, a vowel made by doing nothing but opening the mouth and emitting the first natural sound: "ah." "Ah" was an energy force that came forth from the back of the throat, while vibrating up and down the entire spine. Pointed and direct, "ah" hit the back of the throat like a little hammer of wind.

The next vowel was a longer version of the first, but softer and carrying a quality of release: "aaah." Twice as long as "ah," "aaah" left the back of the throat as if with pleasure: "aaah," that feels good; "aaah," that's nice; "aaah," breathed a sigh of relief. "Aaah" had the quality of traveling up and over, letting go, or giving away. It was gracious and soothing. It erased the clear-cut lines of "ah" into the smooth, rounded edges of "aaah."

Each of these letters had an energetic power created from within the natural contours of the mouth. They were organic shapes, perfected not by holding onto them or by exerting force. They blossomed only through the agency of increased awareness. Surrendering previously learned techniques meant to shape and control, I began to focus, from inside the body, on simply letting go of the sounds. Finding those points where the sounds spoke themselves, I focused more and more intently on those places of natural resonance where the sounds began, tracing them along the body's natural pathways and connecting them with the universal channels of continuous flow.

As we moved from that first position at the back of the mouth into the second position a bit further forward, an "ih" issued forth. I gave up the attempt to associate it with an English sound. I gave up trying to shape it into a perfectly placed Italian vowel. What came through was a simple but powerful being: "ih"–only "ih." Then came the power of "oo." "Oo" was mellifluous, soothing, relaxing, exiting at the front of the mouth. Here, too, I relinquished years of musical training and the attempt to craft "beautiful" tones. I gave up the notion that things could be controlled here or that they would fit neatly into the little boxes of *my* categorized world. Surrendering, I allowed myself to sink into the inverse way.

Having spent a good amount of time discovering the energies of the vowels, we then moved on. Next came the consonants and the mysteries of the breath. The consonants were fully grounded in the body, produced by more solid contact. Some consonants were expressed with small breath, as if held or suspended; others were expressed with big breath, released and allowed to expand. The entire consonantal universe was one of living, breathing sound. The "ka" was held; the "kha" was released. "Ga" was held; "gha" was released. Each consonant was a living force created through one of five points of physical contact found within the mouth. Each, with its own specific character and personality, breathed itself into the world.

Then, combining the pure energy of the vowels into material from the consonantal ground, we moved into the liquid land of the semi-vowels. Vowels, formed from ethereal skies, congealed into matter in the physical realm and what ensued were creatures of true beauty, transformative beyond all words. The unions came: Ya-ra-la-va. Ya-ra-la-va. Ya-ra-la-va. Ya-ra-la-va. Syllables formed like growing seeds, and the sounds of the seeds sprouted deeply within.

When completely engaged in the process of making these sounds, I found that mental distractions fell away. What dropped fell like layers of heavy cement accumulated over years, painstakingly sculpted, patterns accepted or taken on, those given at birth or those given before. But with each new sound came an opportunity to hear and create "nothing but the sound."

"Nothing but the sound" in the first position cut through self-criticism, nourished and cultivated from decades in "school." The second position cleared the judgment of others. "Nothing but the sound" at the third position connected to creativity. The fourth position brought familiar pleasure. And, the power of "nothing but the sound" at the fifth position released centuries of vibration. "Mmm" reached back, recalling a time when words meant what they said, beyond the *Amin* of Muslim mosque, past the *Amen* of early church and ancient temple, coming to rest in the fundamental vibrations of the great, creative *aum*.

Over a time span of just two days, we discovered the sounds of an alphabet whose history is the bedrock in the most ancient core of our being. We combined sounds into words and words into phrases. By the end of our time, we had chanted verses thousands of years old.

We ended the last session with a meditation on the oldest and most sacred *Gayatri mantra*. Chanting to the divine Light at the source of Creation, I focused my spirit and turned within. Listening acutely, I concentrated only on the Light and "nothing but the sound."

Then, as if flowing up from some deep well, a cylinder of silence opened up to echo inside me, grounding me deeply into the earth, and stretching me up and out into the open air. Like a radiating antenna, the cylinder stretched the length of my being, while extending somewhat beyond. I could feel the presence of an opening. Like holy ground. Liquid, flowing luminescence then poured in to fill the void and it lifted me up and out of the room.

I could feel the relative silence in the studio. I could hear the sounds of the city from every direction outside. Focusing just on the sounds, each for a moment, I acknowledged them and then let them go–the wind moving through an open window, a bee bumping against the window pane, the squealing of tires on the pavement and the ensuing blast of a car's horn, a siren wailing in the distance. I concentrated completely on the center of each sound as they blended together to become one long strand carrying me beyond the boundaries of the room. Following the tones, I was a stringless balloon, floating freely on the vibrational ground.

Swells of tone melded together, flooding into my awareness and leaving on the tide. Waves washed into me, and then receded. Floating upward, and further upward, listening simply with inner ears, I heard the serene, steady, distant roar of an ocean striking continuously at some unseen shore. Waves followed one another, billowing upward, rising to crest in swirling white foam, falling to dissipate, exhausted on the shore. Each wave followed in sequence, swelling at first and then retracting within. Rising and swelling, rolling out, only to turn and roll back in again, first expanding and then contracting. This was the water, the ocean of the world. Here was the current, and these are the waves. Now is the time, and this is the way.

I was in the middle of letting go of even the distant, comforting sound of the ocean, so as to slip from the soothing white caps into the velvet silence of blue marine, when I heard our instructor's voice calling us back, asking us to be fully present in the yoga studio. It was only with great difficulty that I summoned my energies to leave the place of peaceful current in order to return to the studio space, which now seemed both familiar and strange. The walls around me were somehow more spacious and the air seemed to reach beyond the room.

I bowed deeply, reverently to the teacher, hoping my eyes would transmit my gratitude, or that the core of my being might communicate what coarse words could not say. I said my goodbyes to my fellow students, radiant links in the sonic chain, and then I left the studio.

A full forty-eight hours after having arrived in the city, I stepped back out into the darkness of the New York night. Unafraid, I didn't question who was behind me. I was quietly aware of everything taking place on all sides. The city's pace had not let up, but I was seeing it in a whole new light. Outlines of buildings hovered in high relief; the pace of a thousand dramas unfolded before my eyes. It was a privilege to be included on this sacred stage. Strains of the old hymn "Jesus Calls Us, O'er the Tumult of Our Life's Wild, Restless Sea" drifted into my awareness, mingling with the afterimages of newly beloved sounds: "*yogash-chitta-vritti-nirodhah.*"

I waited patiently for my train, listening still to the echoes of inner sounds. Finding a seat, I sat in silence. Yet in a state of wonder from the days' events, I glanced down to notice that my usually drumming fingers were still, lying quietly, resting in my lap. My rhythm-driven feet were silent, content to be absorbed in the matter at hand. My entire physical being had come to a point of rest, and the evidence of this was itself visible, seen through the eyes of an inner witness in whose heart profound silence could be felt to resound.

I had been in the presence of strangers but never had I felt more at home. The windows of the train reflected back the disappearing station as we pulled away. My soul had been singed by the flaming tongues of language–white-hot embers smoldered still. The rhythm of the train tracks echoed in the back as we picked up speed. I had stepped out into oceans of silence, into the heat of a thousand suns, beyond the boundaries of the light of sound.

The city was still out there, raging, but I was leaving, now, from a place of peace profound.

☼ ☼ ☼

The events described in this chapter took place during an introductory course on the ancient, sacred language of Sanskrit. It was mentioned above (in Chapter 3) that the ancient Greek philosophers considered music to hold, through harmony and number, the key to the secret mysteries of the universe. Music was held in similar regard in ancient India; seen as a cosmic liberator, "[t]he right kind of music [. . . could] break the cycle of birth, death, and rebirth" (Schneider, *New*, 196). While the Greeks, however, focused on sound and mathematics "as *the* model for true knowing," [the ancient Indians concentrated] on linguistics, that is, on speaking" (Alper, 13). This belief is still easily recognizable in the exalted position with which the Indic culture holds the sacred Sanskrit language, in that, for many, a simple hearing of the sacred words can elicit a "numinous experience" (*ibid.*).

Joseph Campbell has referred to Sanskrit as "the great spiritual language of the world," while the word *samskritam* itself means "well-made," "perfected," "refined," or "polished." The script in which the language is written is *devanaagarii*, meaning "divine abode" or "city of the gods," and the language is said to have been cognized, in deep states of meditation, according to natural cosmic forces, by the ancient poet-sages or *rishis*. "Sanskrit is a language whose harmonic subtlety mysteriously [carries you to source through] the successive phases of creation all the way to origination. This implies the possibility of having speech oriented to a direct living truth" (Houston, 6).

The Indian teachings on sound are many and varied; however, they speak often of the multiple layers of sound and language, layers which range from the coarse outer shell of the spoken word, to the subtleties of mental meaning and intent, to a purer essence of sound, not heard by the physical ear, but perceived by subtler fields of consciousness, sounds found to be aligned with the natural and divine forces of the universe. Sanskrit is a language aligned with those forces, and thus it has been written that every sound in Sanskrit has at least two aspects:

> the more audible sound and the subtler essential sound-element behind, vibrant with the meaning natural to it. This vibrant sense-sound within is the real or fundamental sound, called *sphota*. [. . .] The *sphota* arises in the indivisible, permanent Spirit, and is eternally luminous with power and when the spoken word, its vehicle, is perfectly sounded within and without, it stimulates this inner vibrant activity with the result that the power within responds and illuminates (Tyberg).

Thus, it is said that through chanting the sounds of Sanskrit, and thereby attuning oneself with the natural, luminous vibrancy of creation, one is able to illuminate, balance, and expand one's own consciousness of truth.

Sanskrit is a language specifically developed to bring out various powerful sound vibrations. Every letter in the Sanskrit alphabet has some beautiful, cosmic vibration. That is why each letter can be called a *bijakshara*, or seed word. [. . .] The Sanskrit language, being a perfected one, embodies certain conducive sounds and vibrations that can enrich everyone's life (Bhaskharanada, xi).

Mythological origins of Sanskrit are most often connected with the Goddess. "A Goddess of the Tree was worshipped as early as 2000 B.C. in the Indus valley, where the tree was called the source of all *mantras*—hence, of creation. The tree gave the gift of language to humanity, having a different letter or *mantra* written on each of its leaves" (Walker, 472). In other myths, it is the goddess *Kali* who is credited with inventing the alphabet. Among her many attributes, *Kali* is often portrayed wearing a necklace like a rosary (*japamaala*) of skulls around her neck, and each skull bears one of the forty-nine letters of the Sanskrit alphabet. These letters, the *mantras* of all *mantras*, are called the *maatrikaas* or "the mothers," who bring all things of the universe into being. As she dances, as she speaks, *Kali* chants the Sanskrit letters, which produce the created world. She "wears the letters which She, as Creatrix, bore. She wears the letters which She, as the Dissolving Power, takes to Herself again" (Woodroffe, 238). In this regard, *Kali* is linked to her Vedic "sisters," *Saraswati*, flowing goddess of the powers of music, arts, and knowledge, and *Vaak*, goddess of the creative powers of speech and word.

A similar regard for the mystical power of sound and alphabet is also found within Judaism, as:

"[t]he Hebrew scribe never lost this sense of the letters as living, animate powers. Much of the Kabbalah, the esoteric body of Jewish mysticism, is centered around the conviction that each of the twenty-two letters of the Hebrew alpha-beth is a magic gateway or guide into an entire sphere of existence. [...] The Jewish kabbalists found that the letters, when meditated upon, would continually reveal new secrets; through the process of *tzeruf*, the magical permutation of the letters, the Jewish scribe could bring himself into successively greater states of ecstatic union with the divine." (Abram, 132)

The Kabbalists taught that the power of the Hebrew language could take the chanter, as well as the listener, back to the original wisdom found in the Garden of Eden, where Adam and Eve once walked in innocence. The sounds of the alphabet were portrayed as being "like staircases to the Garden of primordial, pure, and untouched consciousness" (Davis and Mascetti, 179).

The course described in this chapter focused very specifically on the exact placement and pronunciation of the sounds of the Sanskrit alphabet, and the experience of finding and producing those sounds was quite exhilarating in and of itself. However, what further supported and intensified the learning was the method by which the sounds were found and encountered, a method which involved yogic principles of concentration found in the *Yoga-Sutra* of Patanjali (dating somewhere between 200 B.C.E. and 200 A.D.).

It has been said that "[t]he goal of yoga science is to calm the mind, that without distortion it may hear the infallible counsel of the Inner Voice" (Yogananda, 151), and the course was taught using the yogic principles of *abhyaasa* and *vairaagya* to help the student achieve that state of concentration described in the *Yoga-Sutra: yogashchittavrittinirodhah*. This statement is a distillation of the treatise itself and refers to a yogic process of transformation, a process in which there is a quieting of those extraneous activities within consciousness which are not of the Higher Self. The process continues through a series of quietings that encourage transformation, which, in turn, encourage further transformation, until there is a reaching of a realization of the Self.

> According to the Yoga and Sankhya system, sound is the root of all other sensory potentials. It is the sensory quality which belongs to ether, the original element. Hence, through it all the elements can be controlled. The mind itself is composed of sound. It is the reverberation of our words and the ideas they represent which forms the pattern of the mind. Hence, a conscious use of sound both purifies and controls the mind. Most important of the yogas of sound may be the Shabda Brahma Mantra Yoga of Bhartrihari (c. 500 A.D.), which describes the understanding of sound as the highest yoga and the most direct path to the Divine (Frawley in Houston, 14).

The process, then, of using Patanjali's yogic principles (which when applied to sound produces a practice very similar to that of *shabda yoga*) while learning the ancient and sacred sounds of Sanskrit themselves is a profound sonic practice.

CHAPTER 13
SONIC MEDITATIONS –
The Altar of Sound

I suppose the culmination of my sonic investigations came as I began to work with the sounds themselves on a regular basis over a long period of time.

I had been meditating with sound for what seemed like a substantial period. Days had turned into months, and months had become years. What had begun as an intriguing intellectual curiosity had become a spiritual pursuit, a regular daily practice.

Each day I would sit, concentrating my attention on my posture, making sure that my crossed legs formed a stable, triangular base to support the rest of my frame, and that my body was aligned and comfortably balanced, floating gently above its base, spine straight, but not rigid or stiff. I would settle into that comfortable posture, a triangular form, as it created the contours of an inner temple, sacred and refined. In an attitude of worship, I would ready myself for union. Concentrating on the architecture of an inner sanctum, I would open myself to the divine.

Often during meditation, I began by repeating a *mantra* out loud. One such *mantra* had come to me in a dream; it was an ancient chant, and its words were Sanskrit. Forged from the fire of primal energy, the sounds slowed my mental gymnastics, and the content of the words helped me focus on a small inner flame that would appear in the center of my forehead as I closed my eyes. The sounds of the *mantra* made it easier to make the transition from perceiving forms in the physical world as solid, unchanging matter, to understanding them as vital energy forces. Facilitating a shift in awareness, the sounds of the *mantra* were the keys in a lock.

Just as often, I would focus on the syllable "om," periodically repeating the syllable out loud or silently within. Blending my awareness with the sounded vibrations, they were gateways that opened onto other worlds. They were like icons springing alive to the viewer, bestowing grace and transmitting the reality they portray. As symbol became sign, the tones unlocked doors onto layers of vision, bringing worlds of understanding into the conscious realm.

I could sense that my hearing was changing. It was opening up on the inside. A way of knowing was emerging. I was beginning to perceive more through sonic than visual means as layers of sound, previously lodged in the edges of peripheral hearing, came forward, revealing indescribably beautiful visions over time. They presented ancient archetypes from the depths of creation, symbols that have been spoken of, described, and honored on many sacred paths. They conveyed universal understanding, serving as a connector to the deeper awareness that all of life is grounded in sound, and that the truth that lies hidden beneath all things is vibration.

As real as any "real" could be, their depth and detail so convincing, I had to remind myself simply to watch the visions as they arose. Visions presented during meditation can serve as disruptions to deeper, more quieted states. So, just as I watched my breath, inhaling and exhaling, I watched the sonic visions from a place unattached. As the scenes unfolded, rippling forward in ebb and flow, I would watch them and then let them go.

One such archetypal encounter occurred on an occasion during which I had the overwhelming sensation that I had become, on some primordial level, a singing tree with my roots and trunk and limbs all singing different tones, broadcasting melodies out into the world. I had taken my usual posture for meditation, centering myself firmly within its form and had then released the weight of my body into the ground.

I could sense how deeply I was grounded into the earth. "*Lam. Lam. Lam.*" Like long, running roots, my lower limbs reached down and disappeared into the ground. "*Lam. Lam. Lam.*" From the depths came a low, slow whirring, like the rhythms of heavily oscillating blades in the engine of an airplane, or massive thunder rumbling below. "*Lam. Lam. Lam.*" From the subterranean roots emerged a cylinder that formed the outline of a massive trunk, its bark old and rough-grained to the touch. "*Vam. Vam. Vam.*" Stretching upward in elastic strength, its sound was the long, rich tone of a conch shell, spun out over time, sustained. "*Vam. Vam. Vam.*"

In the center of the tree's body, I sensed the smallest, most intensely burning spark. "*Ram. Ram. Ram.*" A tiny jewel, it gave life to the whole tree. "*Ram. Ram. Ram.*" With the sound of a thousand crickets chirping, it divided itself into spinning sparks, each hurling its energy out into the tree's limbs. "*Ram. Ram. Ram.*" Those smaller sparks ignited a brighter flame in the center of the heart, and in the midst of the tree lay a glorious rose, where a soft luxuriance of velvety petals unfolded, uncurling outward in orderly geometric progression. "*Yam. Yam. Yam.*" The petals exhaled as strains of celestial harmony floated mellifluously up into the atmosphere, and broad limbs separated from the trunk, reaching up into the sky.

From there, the wind moved gently through upper branches. "*Ham.*" At the source of the breath, a gateway had opened, and in the throat, sounds pulled the body into alignment as energies crossed into the upper realms. "*Ham.*" While tracing the wind's whistling, there might soon follow the water's sound–a trickling brook or waves gently lapping. "*Om. Om. Om.*" Or the sweetness of a flute might draw me up in a flurry of a thousand peacock plumes with eyes all-seeing and iridescent; a multicolored peacock fan hummed to the drone of a great wind-harp, strings quivering in the wind and shimmering in the upper sunlight. "*Om. Om. Om.*" Yet, somehow I was still firmly rooted in the soil of the dark earth below. "*A-U-M.*"

The sensation of connecting through meditation with this singing tree was accompanied by the knowledge that the tree was very, very old. An ancient being, it stood, its branches arching upward to the skies in a gesture of adoration, along the banks of a great river. Rooted adjacent to ancient waters, and taking its nourishment from them, the tree stood waving in the wind, singing its songs to heal the world. This was the tree from the Garden of Eden; this was the musical tree in Celtic lore. Its name was *ashvattha*; its name was *banyon*. It was called by many names around the world. While adorning the rivers of Revelation, it stood by the waters of Babylon. Birds from all over came to nest in its branches as it sang out its healing songs.

The tree, though ancient, was also accessible. Alive in the body, and attainable through awareness, it was there for all to share. This tree, so ancient yet so familiar, spanned worlds within the living body with tunes grounded in collective memory. The connection lay deep within the body, where hearing was the bridge and sound was the key.

The image of a singing tree that was connected to the body's subtle energy centers was an archetype that came again and again. But other images also came as I practiced the sounds and focused on the words. The vibrations of the *mantras* seemed to bring the images, unfolding glimpses of worlds or beings who lived at the origin of sound–below the words' outer shells heard in the physical world, beyond the words' meanings registered in the mind were worlds of energetic essence centered in the source of the sounds.

One morning, I rose very early. In the pitch–black darkness of the pre–dawn, I took my seat, beginning the ritual of sonic worship. Bowing my head in thanksgiving, I anchored my body's weight into the ground. The night was still and all was quiet as I settled into the sounds.

Focusing intently on where each letter was formed, whether its energy was held or released, I sent the sounds into inner space. Then, like a satellite receiver listening, I waited to receive some communication back. With call and response, I sent and received: chanting and listening, chanting and listening.

Then, in the silence of the night's darkness there issued a phrase "Lift up your hearts," as I heard the words of the *Sursum corda* in a call to prayer. "We lift them to the Lord" answered back, and I was buoyed up by the impulse of the sounds. Expanding upwards, like arms lifting up in prayer, like a flower opening, like a tree in full blossom, like a chalice overflowing, liquid running down its sides, the fullness in my heart engulfed me in sensation, and I was carried up and over a threshold into sacred space.

I could feel the warmth and moisture of a summer night. Crickets chirped everywhere from their hiding places beneath the underbrush. Transfixed in awe, I heard a jingling sound, like the sound of tiny metal pieces, like intricate jewelry or bracelets tinkling as they responded to the body's moves. I sensed a presence leaning over me. Gently enfolding me, Her presence cradled me. A watery rush of smooth sibilance slipped through my inner ears, creating the soothing sounds of Her name, as a warm, protective presence moved through me, from the top of my head down through the bottom of my feet. Embracing me in loving preparation, Her Being washed clear through me, refining, refreshing, cleansing, and blessing.

Then from the breadth of the surrounding sky, a cloak materialized and wrapped itself around my shoulders. It was long and cut from one solid circle, a smaller circle in its center, with an opening fashioned in the front. As it swirled open to gently encompass me, it covered my body from my neck to the ground.

This cloak was substantial, thick, and warm. It was a soft white color, and the material was ancient, heavy, and densely woven. I could see it was trimmed in tiny calligraphic designs. As I looked more closely, I could see the designs were letters. The cloak was trimmed in tiny letters, running in a circle around the collar, and encompassing the entire length of the hem while running in straight rows down the front and the back.

The letters were small, living beings. They had dimension and movement, shape and form. Each one of them was a power of light, an energy radiating a distinctive pattern. Some were made of cooling candescence, while others were full of iridescent fire. Some were shades of rich purple or green; others were blazing reds or ice-cooled blues. Each of the letters broadcast a particular quality of sound into the world, and as I focused intently on one letter after the other, I could see how their patterns fit together. The blending of certain letters made sounds soft and smooth, while other combinations created rhythmic outbursts jagged and quick.

This robe of light radiated the energies of creation, and when chanting, the singer became aligned with the ancient forces that sustained the universe. The cloak conveyed an indescribable feeling of joy and the clear perception of truth. It was protection; it was reassurance; it was a way of knowing, direct from the depths of the creative word.

There, deep in the ether, I sensed a deeper presence of an immense and sacred form. Shaped by features ever changing, it was difficult to fix on, as it was continually being transformed. At first, it was an oblong, marble table, candle flame hanging high overhead. Then it grew into more of a square, in a pit made of bricks, sunken in the ground where fires glowed. From there it moved to form a colossal circle of rocks that ringed a center. At times, its shape disappeared altogether or it took on the organic shape of a great bird.

On the one hand, this changing sacred space seemed to resonate from someplace deep inside myself. It was as if each time I sang or chanted, an abyss was crossed with personal prayer. I could feel the individual elements that had come together to create "me": the patience and endurance that were my mother; the tradition of the spoken word that was my father; the impatience and curiosity that were me; the many traditions of sound and music in all my teachers' songs. All the individual elements of particularity were offered here in loving worship in this interior form.

But this was no earthly altar. It did not belong only to me. It reached far beyond any *one* individual to a place outside the physical plane. Born in the ether, and universal, it stretched out between the ears and mouth of divinity. It held a holy space where deity could hear and respond. This was the place from which creation had been spoken, as well as the place to which all would return. This wide-open space of blue-to-white vastness was a place of hearing and a place of sound; it was a holy space of chanting and listening. This was the source of life; this was *the* altar of sound.

Words of prayer and enchanted melodies formed its foundation. Each musical strain, a building block, interlocked with the others to form a continually modulating structure of sound. Sounds here reverberated forever and the melodies were ages old. Songs chanted centuries–even millennia–ago gathered from places around the globe as the essence of their pure sonic forms mingled in color-filled streams and ribbons of light. Songs, offered once as prayer–chanted, walked, whirled, or sung–merged here in luminous form, transparent as mist, in dazzling array.

Here, the sacred traditions mixed with ease. Strains of *"Kyrie Eleison"* and *"O Oriens,"* Gregorian melody offered from the Christian West, fused freely with the *Gayatri mantra* born in the Vedic East. Chants of the Buddhist tradition, *"Gate, gate, para gate…"* harmonized with *Shema Israel,*". . .the Lord thy God is one." Native drummings proclaiming *mitakuye oyasin,* "we are all related," mingled with Sufi rhythms of *zikr* offered to Allah. The sacred vowels of ancient Egypt blended with the majesty of Druid tones. "Aye-eye-oh" harmonized with the great *Al-le-lu-i-a,* while *Amen* and then "om" under-girded them all. On this level of sonic essence, this sacred level of worship and prayer, all were related.

Over and around this changing altar, bands of color continually moved. Each of the bands was, in its essence, a sonic ribbon, and each ribbon spun out long, vibrant threads, like the wind sending strands of enchanting perfume. The sonic threads mixed and intermingled, creating an aural fabric in gossamer sheets of billowing sound. This material contained every hue, shimmering as it moved, while a clear, white crystal substance lay at its core.

As the sonic threads blended with each other, layers of harmonics hummed. Subtle sounds oscillated back and forth frequently, resonating at the octave, then singing at the fifth. These tones of celestial harmony radiated streams of light, as yards of humming sonic fabric stretched out as far as the eye could see. Mile on mile of creative matter reached out into every corner of the world, making all that could be seen, seen, and all that could be heard, heard.

Eventually, at the far reaches of the fabric, all the sounds came together to curl in on themselves. Dancing rhythmically into a great vortex, they moved in time, becoming a massive pillar of swirling sound. All the chords, every meter, each strain disappeared into the centripetal pull of that colossal column, and as each sound evaporated into the spiral, the melodies quieted and the vibrations ceased. At the center of the massive pillar, everything circled into silence, all came to rest in quiet and calm.

> Then . . . peace . . . quiet . . . silence pervaded. Echoes hushed.
>
> Nothing moved.
>
> Time's rhythms halted, and all was still.
>
> Silence reigned without end.
> Then, from somewhere, I heard a rush, like wind

whooshing in through a long tunnel, or like a breath being suddenly drawn. Flooding in, the sounds returned, and I came again to a new awareness, sitting in the darkness of the early morning, just before the break of dawn:

> Inhaling in, I could sense the altar. Exhaling out, I could feel its forms.
>
> Inhaling in, I could see its colors. Exhaling out, I could hear its sounds.
>
> Breathing in, I could translate the light now.
>
> Chanting out, I could sing the songs.

※ ※ ※

This last chapter begins where the first one ended, by focusing on the breath, and by working with sound in a contemplative manner. In a way, a circle has been completed in that innate behavior from childhood, lost for years, has been reclaimed in a regular musico-spiritual practice. This practice involved meditation with sound and the use of music as a creator of sacred space. It is a practice of sacred singing, connected with an understanding of sound as being a manifestation of Primal Music.

Sacred singing is centered in the breath: "Primal Music comes directly from the Breath of Life, [. . .] which gives birth to the voice [and can be heard in certain] types of song or chant [from] Bhuddist *mantra*, [to] Islamic prayer call [to] Christian plainchant" (Stewart, 52). Awareness of the breath in Primal Music is an act of worship itself: "Breath is [. . .] regarded, as early as Vedic times, as having a ritualistic significance [in that an inner worship occurs] which consists of offering the breath as an oblation in speech (when one is speaking) and speech in breath (when one falls silent . . .); that is to say that it considers both functions as ongoing oblations. In this, of course, the body is assimilated to the sacrificial altar, and the adept's life itself becomes a sacrifice" (Padoux, 26, n. 59).

Just as "sound sacrifice" is grounded in the breath, sacred singing is centered in the voice: "A sound sacrifice is accomplished by a singer who surrenders his breath and individuality in order to bring expression to an impersonal empty form and pushes his own personality into the background in order to make himself the sounding board of a higher principle (Schneider, "Acoustic," 75). Sacred singing encompasses a spiritual journey from the physical music to the unstruck sound of the soul: "The sound produced by sacrifice [. . .] establishes the connection between heaven and earth" (Schneider, *New*, 49), and the singer becomes a bridge between the worlds.

Sacred singing is a play of relationships, hearing and listening, chanting and listening, sound and silence: "The value of music is that it links the physical and metaphysical together in the human consciousness, creating modes of awareness in which the creative unity of all being is glimpsed, albeit briefly" (Stewart, 53). And, as sacred relationship grounded in worship, the way that sound is produced is as important as the sound itself: "The performers of these primordial [. . .] sacred chants paid as much attention to the way a tone was reached as to the tone itself–just as a true lover considers the way he or she approaches the beloved as important as the act of love itself" (Rudhyar, 14).

Sacred singing is a celebration of the God-given gift to praise and to create through the voice: "The Sanskrit word for 'musician' is *bhagavathar*, [s]he who sings the praises of God [and] great religious music of East and West bestows joy on man because it causes a temporary vibratory awakening [and] a dim memory comes to him of his divine origin" (Yogananda, 157).

This kind of song sacrifice is about inner worship, not about public performance: "The vibratory essence of sound affects the inner walls of the nerves and the blood vessels. The inner walls of each cell resonate and the power of vibration (sound) affects not only the physical cell. [. . .] Chanting implants in the psyche the basis for the new and fine-tunes the physical body for both spiritual and mental growth" (Rael, 121).

This kind of singing is the realization of the divine within the ordinary, placing the singer, through these acts of inner reverence and worship, in touch with the very energies embodied in the creative songs being sung: "The plunge into the depths of consciousness–with mind completely stilled and in a poised, receptive state of awareness, results in revelation. Such revelation or inner seeing may take the form of vision, of sudden flashes and realizations of great truths otherwise left unconceived or of communion with denizens of another dimension of life, or their manifestation" (Miller in Feuerstein, *In Search*, 188).

This kind of singing allows a shift in perception, for as Don Campbell writes so beautifully in *The Roar of Silence:* "To sing is not enough. We must tap the powers that lie beneath the consciousness of the song. The song leads down to the chant, the chant leads down to the tone, the tone leads down to the breath, and the breath leads to the energy beneath the sound–the roar of silence" (Campbell, Roar, 75–76).

In sacred singing, the power of *mantra* reaches beyond the realm of ordinary words, as it is felt that *mantra* holds a deeper power to create. "Claimed as *man-tra* (*min*–d ins–*tru*–*ment*), [. . .] it is said to have the ability to re–create in the prepared hearer the experience of the poet, of the *rishi*. Thus, the word is not just a sound arbitrarily connected with an object or event, but is, essentially, a voice, a force producing an effect directly on the substance of being. It is a creative, living symbol" (LeMee, xi), and its power "lies in its ability to create new vibrational patterns that through resonance can stabilize our mental attitudes and physical energies" (McClellan, 65).

Mantra is sacred prayer; it is sacred song that "gradually converts the devotee into a living center of spiritual vibration which is attuned to some center of spiritual vibration vastly more powerful than his own," and the spiritual effects produced through this vibrational attunement are primarily those of love (Twitchell, 82-3). The best known of all *mantras* is probably the *mantra* "aum," or "om." The sound *A-U-M* is comprised of the first and last vowel sounds of the Sanskrit alphabet, along with the last and most resonant of the consonants (*M*); this sound, therefore, encompasses all of the sounds capable of being produced within the human mouth. As such, it is both literally all sounds, and a symbol of All-Sound, All-Sound being the emanation of God as Creator. It is a "universal symbol-word for God, [. . .] the all-pervading sound emanating from the [. . .] Invisible Cosmic Vibration [. . .] *Aum* of the Vedas became the sacred word *Hum* of the Tibetans, *Amin* of the Moslems; and *Amen* of the Egyptians, Greeks, Romans, Jews, and Christians" (Yogananda in Cornell, 56).

The use of "om" (or of any *mantra*, for that matter) has been referred to as a boat, as the words of the mantra help the chanter to cross the "waters of this world" and attain the spiritual "other shore." "The *om* sound is truly liberating because it expands the reciter beyond the physical boundaries, thus, restoring the recognition of the universal Self as his or her true identity" (Feuerstein, "The Sacred," 6).

Those archetypal images that arose during this time of meditation (and I emphasize they occurred over an extended period of time, the story being a condensation of several events) are interesting in that some of them derived from symbols I can directly connect to my own life–images of the chalice and of the altar, of course, being connected with priesthood–whereas others were of a more universal nature, arising from some place apparently deeper in my sonic awareness. Even more interesting, however, is the fact that I identified the images as being "universal" only when reading about them in written sources *after* I had encountered them in meditation.

So, while I was one individual experiencing the images in the story, they are images of universal occurrence, and are in no way meant to suggest unusual or special events–only that such occurrences do take place within the inner realms of consciousness. Such images included: the subtle energy centers of the body (the *chakras* and their sounds), a singing tree (connected with the *chakras*, and found in many ancient, spiritual mythologies in differing forms), a crystal pillar (found in Celtic mythology, or in the Hebraic "pillar of cloud"), a sonic cape (found in Native American myths of creation, as well as in Psalm 104 where "divinity comes clothed in a robe of light"), and the idea of a cosmic sonic fabric (found in various Native American descriptions of the web-of-all-relations, as well as in the Vedic descriptions of the sacred singer who creates the universe through his or her sacrificial singing and the loom of the mouth). Likewise, the realm of the pre-dawn, that time-frame in which this story takes place, is referred to in many traditional mythologies as the time of the "singing light," as that magical time of worship when creation is traditionally sung or breathed into being. As the Psalmist cries: "Awake, psalteries and harps! I will awaken the dawn!" (Psalms 57:9).

EPILOGUE

Isaiah must have trembled, "in the year that King Uzziah died," as he stood or sat within the temple, encountering a presence so vast and so powerful that he shrank from it, consumed by thoughts of his own insignificance and fears of his own shortcomings. Yet, deep within the actual encounter, there burned a power, an energy which, formed in the image of burning-hot coals carried by the seraphim to the prophet's mouth, cleansed a reluctant Isaiah and cleared a way for him to speak of profound and mysterious things.

It is with Isaiah's sense of awe, as well as his feelings of unworthiness, that I have been perceiving something hidden within the world and within the world's sacred teachings–a sonic thread–a living, spinning, vibratory thread that connects all life to all life. It is spoken of by every one of the world's religious traditions, yet we remain deaf to each other's prayers. It sings from the center of every spiritual pathway, yet our ears are stopped; we do not sing each other's songs. This sonic thread is so wonderfully powerful, so all-encompassing, that one's first inclination is not to write or speak of it at all. In comparison to one's abilities, it is far too vast, much too holy. And yet, through actual encounter with that sacred thread has come a kind of preparation from which one prays the right words will be given, and in which one trusts an expression will come that will be worthy of the beauty and the depth of the world's light-giving sacred sounds.

Isaiah called to the people to hear God's voice and to listen once again to sacred sound. Through the medium of sound and the agency of the human voice, Isaiah asked the people to unblock their ears and restore their relationship to Holiness, the central characteristic of which was marked by the exchange that takes place in the act of deep listening.

I am no Isaiah. But I live in a time when it would seem that we, too, "hear and hear, and do not understand," shutting out our capacity to recognize sacred sound, dulled by the roar of the sports arena, the constant drone of radio and TV, and the deafening rush of supersonic flight. Almost three thousand years after Isaiah, it would seem that we have only improved our ability to turn a deaf ear and to use our God-given capacity for listening–if we acknowledge it at all–for little more than entertainment or pleasure. We, too, deadened by the frantic pace of extroverted activity, suffer from blocked ears and silenced song.

In the midst of the busy actions of service, I came full circle into devotion again. Parched by the demands of activity, a still, small voice pulled me from constant motion into a silent retreat for meditation and into the depths of chanted prayer.

There I found refreshment in universal waters. As I moved my focus from exterior activity to interior worship, I tasted what the world's spiritual traditions had to offer and I found new life through images in sound. Each tradition had such beauty: from the Native American web that hums creation to the singing tree of the ancient Celts, from the Vedic waters of the world issuing forth from the mouth of Diety, where *mantra* is the boat to traverse the waves, to the Sufi *zikr* where movements are reminiscent of Dante's celestial dance. Each of the traditions spoke so powerfully, while each perceived sound in a similar vein. Sound was creation and continued sustenance. The foundation of community, sound brought balance through story, infusing new life through the memories of song. Bringer of harmony, sound was medicine for the world's woes. Sound was the Word. Sound was a savior. Sound was a peacemaker. Sound was the Way.

From the inquiries that followed from that simple retreat came a whole new understanding of music. Beyond the scope of entertainment, sound holds the power to heal and the ability to transform. Its effects can be felt in the body of the individual, bringing about balance through an attention to one's own inner tunes. Its effects can be felt in the joining of religious communities, as a single, sonic current undergirds them all. Its effects can be felt in the world, as the traditions of prayer and song call us to work together, to care for our world, and to live in peace.

However, these effects are most profoundly felt because, while bringing balance and restoring harmony, sound has the fundamental ability to bring us back to an original awareness, to return us back to original source. For, above all, sound is Source. And, therein lies its power; therefrom flows the song.

Whether anyone will recognize the sonic thread, or comprehend the altar of sound, is hard to say—everyone's path is different. I can only explain that my sonic investigations have resulted, for me, in a return to Source. Through the avenues of sound and the agencies of chant, I have experienced, time and again, that place of life-giving transcendence where "eventually, all things merge into one" in a river of sound that rushes over the rocks of time and under which the words of creation are written. It is a place where Norman Maclean found he had become "haunted by waters," a place where I, too, like someone enchanted by memory, will continue to return, to splash in the reverie of the words and to drink in the beauty of the sounds.

۞ ۞ ۞

Originally, I did not plan to include short stories in this book. I had intended to write a book about the spirituality of music and the importance of sound in the spiritual life. It was to be a wide-ranging book that would document the use of sound and music in spiritual practices around the world, and one that would investigate some of the sonic similarities within those traditions. I was working on the "harmonic hunch" that if I listened deeply enough, I would find common threads concerning the spirituality of sound within the world's great religious traditions.

After immersing myself in study and direct exploration, I came to a point where what I was hearing had become overwhelming. While elated to find that my hunch had been very definitely confirmed, I was completely inundated by volumes of literature, the vast richness of many spiritual traditions, and the stunning sonic similarities. Meanwhile, I was finding that the sounds of the spiritual practices themselves were opening up new avenues of awareness within my own consciousness. The sounds were working on me, changing me, at the level of experience. I found that I had become involved in much more than a research project, as I began to realize that the sounds themselves were spiritually initiating me.

Since this study had brought such refreshment and joy directly to me, I decided to write the book as a group of short stories that might capture some of the feelings and the life experience, while still depicting different aspects or traditions of sound and spirituality. The stories portray concepts and traditions, while the notes are meant to elucidate meaning and context, and to suggest additional reading. These stories are not an academic treatise, nor are they able, in the short space of thirteen stories, to speak of all the beautiful and profound traditions that use sound in spiritual or transcendental ways.

They are stories that suggest sound as a spiritual path, and put forth the pursuit of personal and global harmony as a spiritual goal. Sound practices are transformational. From singing lessons to mantric chanting, sound practices are balancing and harmonizing. They touch our spirits and connect us with one another. Sacred sound is a spiritual peacemaker. Beyond the realm of religious doctrine, it calls us to recognize a deeper spiritual commonality of sound in the realms of the heart.

> At the heart of each of us, whatever our imperfections, there exists a silent pulse of perfect rhythm, a complex of wave forms and resonances, which is absolutely individual and unique, and yet which connects us to everything in the universe. The act of getting in touch with this pulse can transform our personal experience and in some way alter the world around us.
> (Leonard, xii).

Meditation and the practices of sound meditation are not isolated, solitary acts, as some may argue. Every tradition mentioned in this book teaches that being in touch with the inner harmonies helps us to be more responsive to the world around us, to be more compassionate, to be more loving. For while experiencing transcendence through sound and meditation, we experience "a new capacity for true empathy, a capacity to be at peace with others, and indeed at peace with the whole of creation" (Main, 13).

In the Christian tradition, it has been written that *mantra* leads us to an awakening of the Holy Spirit "the essential Christian experience, the experience of being born again in the Holy Spirit [. . .] when we realize the power of the living Spirit of God within us [. . .] that God's love has flooded our inmost heart through the Holy Spirit He has given us" (*ibid.*, 16). In the Jewish tradition, it has been written: "When we meditate we naturally expand and cultivate *Chesed*–compassion and loving-kindness. The function of meditation within Judaic mysticism is not solely to improve one individual; the ancient prophets quoted in the Bible were engaged in *tikkun olam*, healing of the world" (Davis and Mascetti, 131). By connecting our own woundedness to

God, we become bearers, in our own way, of healing to those in our families, those in our communities. Prophets on a small scale, we become bearers of *tikkun olam* (*ibid.*).

From the Sikh tradition, it has been written: "The mystic after his mystical experience returns to life much richer, much more powerful, much more effectively co-operant, much more generous, liberal, much more love-serving and much more sympathetic and knowing" (Singh, 23). And, from the Sufi tradition, it has been written: "music has a mission [for individuals and for the multitudes. . . .] All the trouble in the world comes from lack of harmony. [. . .] So if the musician understands this, his customer is the whole world" (Khan, *Music of Life*, 63).

We are the instruments of attunement. Each person, tuning himself or herself inwardly, can, through these practices of harmony, help to tune others and the world. Far from being exercises of self-absorption that isolate one from the world, sound meditation "can put an end to compulsive attachment; it can wean you away from serious addictions; it can reduce your self-will; it can bring great security to your mind" (Easwaran 208). The use of *mantra* clarifies consciousness, releasing fear and negativity, and as concentration becomes a natural state, "we can turn our attention to any problem and penetrate to the heart of it. [. . .] Now, however, we will see only the unity of life, and all our energy will be directed to solving the biggest problems we face today–violence, the despoliation of the environment, the disintegration of the family" (*ibid.*, 213).

Furthermore, sound practices and meditation can have a direct effect on our health, from the physical health of the individual to the spiritual health of the world. In his ground-breaking work, *The Mozart Effect*, Don Campbell explains how each of us "can use music and self-generated tones to help us become more sensitive to our own rhythms and cycles" (Campbel, Mozart, 121). He offers a wonderful array of stories about how people have used the powerful medium of sound to heal themselves.

The very language we use to describe the state of being healthy resounds with sonic images:

> Hidden within our language are metaphors which describe the connections between music and health. [. . .] To be alive means to have a pulse. To be "sound" means to be healthy. To be well is to be in harmony with one's surrounding, to be "in tune." When our muscles are strong, they are well-toned. When we are sick we are "out-of-synch." All of these are musical terms (Hale, 94).

The very basis of what we believe it means to be physically healthy is grounded in sound. Likewise, the basis of mental and spiritual health is grounded in sound. "In Latin, the term meaning 'to sound through something' is *personare*. Thus, at the basis of the concept of a *person* (the concept of that which really makes a human being an unmistakable, singular *per-sonality*) stands a concept of sound: 'through the tone'" (Berendt, *The World*, 171). Sound resonates from the very depths of personhood. And, the degree to which each *per-son* is in tune with the higher vibrations of universal concern, the greater is the extent to which that *per-son* can have a harmonizing effect on his or her surroundings.

In 1969, Corrine Heline wrote: "The time is fast approaching when people will select their music with the same intelligent care and knowledge they now use to select their food. When that time comes, music will become a principal source of healing for many individual and social ills, and human evolution will be tremendously accelerated" (Heline, *Esoteric*, in Halpern and Savary, 150). If that prediction comes true, perhaps we will have succeeded in re-tuning ourselves by returning to those teachings that the ancients knew so well. Music will no longer be a pastime, or a way to forget about God, but music and sound will have become again a way to realize God, a path for "spiritual attainment and healing of the soul" (Khan, *The Mysticism* in Halpern, 180).

The stories are experiential and admittedly autobiographical, but they aim to portray events that are far greater than my own. Though they focus on experiences I have had while working with various sound practices, those experiences are universal, and are meant to be a springboard to a greater understanding of those practices in which, through greater knowledge and exposure, everyone can participate.

It is my hope that the medium of story will move readers more immediately to recognize the importance of sound and harmony in their own lives. The stories are a few threads of melody plucked from the Altar of Sound, intended to encourage the readers to listen more deeply and to attune themselves to the melodies of the Celestial Song that are sounding within their own souls.

BIBLIOGRAPHY & FURTHER READING

Chapter 1

Abram, David. "The Forgetting and Remembering of the Air," in *The Spell of the Sensuous: Perception and Language in a More-than-Human World*. New York: Pantheon Books, 1996.

Campbell, Don. *Introduction to the Musical Brain*. St.Louis, Missouri: Magnamusic–Baton, 1984.

Campbell, Don G. *Music: Physicians for Times to Come*. Wheaton, IL: Quest, 1991.

Chowdry, Dr. L. R. *Pranic Healing*. New Delhi: B. Jain Publishers, 1997.

Goldman, Jonathan. *Healing Sounds*. Rockport, Maine: Element Books, 1992.

Hale, Susan Elizabeth. *Song and Silence: Voicing the Soul*. Albuquerque: La Alameda Press, 1995.

Katsh, Shelley and Carole Merle–Fishman. *The Music Within You*. New York: Simon and Schuster, 1985.

Kramer, Kenneth. *The Sacred Art of Dying: How World Religions Understand Death*. Mahwah, New Jersey: Paulist Press, 1988.

Rael, Joseph. *Being and Vibration*. Tulsa: Council Oak Books, 1993.

Roth, Nancy. *The Breath of God: An Approach to Prayer*. Cambridge, Massachusetts: Cowley Publications, 1990.

Sivapriyananda, Swami. *Secret Power of Tantrik Breathing*. Delhi: Abhinav Publications, 1996.

Weeks, Bradford, M.D. "The Physician, the Ear and Sacred Music," in Don Campbell, *Music: Physician for Times to Come*. Wheaton, Illinois: Quest Books, 1991.

Wilson, Tim. "Chant: The Healing Powers of Voice and Ear," in Don Campbell, *Music: Physician for Times to Come*. Wheaton, Illinois: Quest Books, 1991.

Zi, Nancy. *The Art of Breathing*. New York: Bantam Books, 1986.

Chapter 2

Abram, David. *The Spell of the Sensuous: Perception and Language in a More-than-Human World*. New York: Pantheon Books, 1996.

Crandall, Joanne. *Self-Transformation through Music*. Wheaton, Illinois: Theosophical Publishing, 1988.

David, William. *The Harmonics of Sound, Color and Vibration*. Marina del Rey, California: DeVorss, 1988.

Easwaran, Eknath. *The Unstruck Bell*. Tomales, California: Nilgiri Press, 1993.

Erndl, Kathleen M. *Victory to the Mother*. Oxford: Oxford University Press, 1993.

Ferry, Judith. "Sanskrit Training for Professionals," *Sanskrit Today*, Fall 1994, 1–2.

Feuerstein, Georg, Subhash Kak and David Frawley. *In Search of the Cradle of Civilization*. Wheaton, Illinois: Quest Books, 1995.

Gardner, Kay. *Sounding the Inner Landscape: Music as Medicine*. Stonington, Maine: Caduceus Publications, 1990.

Halpern, Steven. *Sound Health: The Music and Sounds that Make Us Whole*. San Francisco: Harper & Row, 1985.

Khan, Hazrat Inayat. *The Music of Life*. New Lebanon, New York: Omega, 1983.

McAllester, David P. "Coyote's Song," *Parabola*, Vol. V No. 2, 47-54.

Roth, Nancy. *The Breath of God: An Approach to Prayer*. Cambridge, Massachusetts: Cowley, 1990.

Schneider, Marius, "Acoustic Symbolism," in *Cosmic Music*, Joscelyn Godwin, Ed. Rochester, Vermont: Inner Traditions, 1989.

Chapter 3

Cirlot, J.E. *A Dictionary of Symbols*. New York: Philosophical Library, 1971.

Crandall, Joanne. *Self-Transformation through Music*. Wheaton, Illinois: Theosophical Publishing, 1988.

Danielou, Alain. *Music and the Power of Sound: The Influence of Tuning and Interval on Consciousness*. Rochester, Vermont: Inner Traditions, repr. 1995.

Ficino, Marsilio. *The Book of Life*. Charles Boer, trans. Dallas: Spring Publications, 1980.

Fideler, David. "Orpheus and the Mysteries of Harmony," *Gnosis Magazine*, No. 27, Spring 1993, 21-27.

Flebotte, Thomas R. and Mario A. Ortiz. "The Transcendent Harmony: Celestial Music in The Paradiso of Dante," *Sophia*, II/2 (Winter 1996), 46-80.

Godwin, Joscelyn. *Music, Mysticism, and Magic: A Sourcebook*. New York: Arkana, 1987.

Godwin, Joscelyn. *The Harmony of the Spheres: A Sourcebook of the Pythagorean Tradition in Music*. Rochester, Vermont: Inner Traditions, 1993.

Khan, Hazrat Inayat. *Music*. Claremont, California: Hunter House, 1988.

Kristeller, Paul O. *The Philosophy of Marsilio Ficino*. Gloucester, Massachusetts: Peter Smith, 1964.

Claude V. Palisca. *Humanism in Italian Renaissance Musical Thought*. New Haven, Connecticut: Yale University Press, 1985.

McClellan, Randall. *The Healing Forces of Music: History, Theory, and Practice*. Rockport, Massachusetts: Element Books, 1991.

Meyer-Baer, Kathi. *Music of the Spheres and the Dance of Death*. Princeton: Princeton University Press, 1970.

Plato. *The Republic*. London: Penguin, 1987.

Rice, Eugene F., Jr. *The Renaissance Idea of Wisdom*. Cambridge, Massachusetts: Harvard University Press, 1958.

Tame, David. *The Secret Power of Music: The Transformation of Self and Society through Musical Energy*. Rochester, Vermont: Destiny Books, 1984.

Trinkhaus, Charles. *In Our Image and Likeness: Humanity and Divinity in Italian Humanist Thought*. Chicago: University of Chicago Press, 1970.

Walker, D.P. *Spiritual and Demonic Magic from Ficino to Campanella*. Liechtenstein: Kraus Reprint, 1969.

Yates, Frances A. *Giordano Bruno and the Hermetic Tradition*. Chicago: University of Chicago Press, 1964.

Chapter 4

Alper, Harvey P., ed. *Understanding Mantras*. Albany, New York: SUNY Press, 1989.

Campbell, Don G. *Music: Physicians for Times to Come*. Wheaton, IL: Quest, 1991.

Crandall, Joanne. *Self-Transformation through Music*. Wheaton, Illinois: Theosophical Publishing, 1988.

Dange, Sindhu S. *Aspects of Speech in Vedic Ritual*. New Delhi: Aryan Books, 1996.

Feld, Stephen. *Sound and Sentiment: Birds, Weeping, Poetics, and Song in Kaluli Expression*. Philadelphia: University of Pennsylvania Press, 1982.

Feuerstein, Georg, Subhash Kak and David Frawley. *In Search of the Cradle of Civilization*. Wheaton, Illinois: Quest Books, 1995.

Gardner, Kay. *Sounding the Inner Landscape: Music as Medicine*. Stonington, Maine: Caduceus Publications, 1990.

Gardner-Gordon, Joy. *The Healing Voice*. Freedom, California: The Crossing Press, 1993.

Gillette, Douglas. *The Shaman's Secret*. New York: Bantam Books, 1997.

Godwin, Joscelyn. *Harmonies of Heaven and Earth*. Rochester, Vermont: Inner Traditions, 1987.

Godwin, Joscelyn. *The Mystery of the Seven Vowels*. Grand Rapids, Michigan: Phanes Press, 1991.

Goldman, Jonathan. *Healing Sounds*. Rockport, Massachusetts: Element Books, 1992.

Hale, Susan Elizabeth. *Song and Silence: Voicing the Soul.* Albuquerque: La Alameda Press, 1995.

Halpern, Steven. *Sound Health: The Music and Sounds that Make Us Whole.* San Francisco: Harper & Row, 1985.

Hamel, Peter Michael. *Through Music to the Self.* Longmead, Dorset: Element Books, 1991.

Khan, Hazrat Inayat. *The Music of Life.* New Lebanon, New York: Omega, 1983.

Keyes, Laurel Elizabeth. *Toning: The Creative Power of the Voice.* Marina del Rey, California: DeVorss, 1973.

Mahlberg, Arden. "Getting the Ego Humming," in Don Campbell, ed. *Music and Miracles*, Wheaton, Illinois: Quest Books 1992, 219–29.

Newham, Paul. *The Singing Cure.* Boston: Shambhala, 1994.

Padoux, Andre. *Vac: The Concept of the Word in Selected Hindu Tantras.* Albany, New York: SUNY Press, 1990.

Schneider, Marius, "Primitive Music," in Egon Wellesz, ed. *The New Oxford History of Music: Vol. I Ancient and Oriental Music*, London: Oxford University Press, 1957.

Staal, Frits. *Ritual and Mantras: Rules without Meaning.* Delhi: Motilal Banarsidass, 1996.

Stewart, R.J. "Vowel Sounds and Music," in *The Spiritual Dimension of Music.* Rochester, Vermont: Destiny Books, 1990.

Weeks, Bradford, M.D. "The Physician, the Ear and Sacred Music," in Don Campbell, *Music: Physician for Times to Come.* Wheaton, Illinois: Quest.

Yogananda, Paramahansa. *Autobiography of a Yogi*. Bombay: Kothari, repr. 1994.

Chapter 5

Abram, David. *The Spell of the Sensuous: Perception and Language in a More-than-Human World*. New York: Pantheon Books, 1996.

Achtemeier, Paul J., ed. *Harper's Bible Dictionary*. San Francisco: Harper and Row, 1985.

Aldredge–Clanton, Jann. *In Search of the Christ-Sophia*. Mystic, Connecticut: 23rd Publications, 1995.

Easwaran, Eknath. *The Unstruck Bell*. Tomales, California: Nilgiri Press, 1993.

Fideler, David. *Jesus Christ: Sun of God*. Wheaton, Illinois: Quest Books, 1993.

Fideler, David. "Orpheus and the Mysteries of Harmony," *Gnosis Magazine*, No. 27, Spring 1993, 20–27.

Gawronski, Raymond. *Word and Silence*. Grand Rapids, Michigan: Eerdman's, 1995.

Granville, C. Henry, Jr. "Word and Wisdom," in *Logos: Mathematics and Christian Theology*. Cranbury, New Jersey: Associated University Press, 1976.

Johari, Harish. *Tools for Tantra*. Rochester, Vermont: Destiny, 1986.

Khan, Hazrat Inayat. *The Music of Life*. New Lebanon, New York: Omega, 1983.

Kinsley, David. *Hindu Goddesses*. Berkeley, California: University of California Press, 1986.

Padoux, Andre. *Vac: The Concept of the Word in Selected Hindu Tantras.* Albany, New York: SUNY Press, 1990.

Rael, Joseph. *Being and Vibration.* Tulsa: Council Oak Books, 1993.

Singh, Sawan. *Philosophy of the Masters.* Vol IV. Punjab, India: Radha Soami Satsang Beas, 1967.

Walker, Barbara. *The Woman's Dictionary of Symbols and Sacred Objects.* San Francisco: Harper, 1988.

Winston, David. *Logos and Mystical Theology in Philo of Alexandria.* Cincinnati: Hebrew Union College Press, 1985.

Chapter 6

Apel, Willi. *Gregorian Chant.* Bloomington, Indiana: Indiana University Press, 1990.

Benedictines of Solesmes, eds. *The Liber Usualis.* New York: Desclee Co., 1962.

Bourgeault, Cynthia. "The Hidden Wisdom of Psalmody," *Gnosis Magazine,* Fall 1995, 22–28.

Campbell, Don G. *The Mozart Effect: Tapping the Power of Music to Heal the Body, Strengthen the Mind, and Unlock the Spirit.* New York: Avon Books, 1997.

Campbell, Don G. *The Roar of Silence: Healing Powers of Breath, Tone and Music.* Wheaton, IL: Quest 1989l

Del Bene, Ron. *Into the Light: A Simple Way to Pray with the Sick and the Dying.* Nashville: The Upper Room, 1988.

Evola, Julius. *The Yoga of Power.* Rochester, Vermont: Inner Traditions, 1992.

Filippi, Gian Giuseppe. *Mrtyu: Concept of Death in Indian Traditions*. New Delhi: DK Printworld, 1996.

Findly, Ellison Banks. "Mantra kavisasta," in Harvey P. Alper, ed., *Understanding Mantra*. Albany, New York: SUNY Press, 1989.

Fremantle, Francesca and Chogyam Trumgpa. *The Tibetan Book of the Dead: The Great Liberation through Hearing in the Bardo*. Boston: Shambhala, 1975.

Hale, Susan Elizabeth. *Song and Silence: Voicing the Soul*. Albuquerque: La Alameda Press, 1995.

Kramer, Kenneth. *The Sacred Art of Dying: How World Religions Understand Death*. Mahwah, New Jersey: Paulist Press, 1988.

LeMee, Katharine. *Chant: The Origins, Form, Practice, and Healing Power of Gregorian Chant*. New York: Bell Tower, 1994.

Main, John. *Word into Silence*. New York: Paulist Press, 1981.

Meyer-Baer, Kathi. *Music of the Spheres and the Dance of Death*. Princeton: Princeton University Press, 1970.

Nugent, Christopher. "Illumination: Nocturne," in *Mysticism, Death, and Dying*. Albany, New York: SUNY Press, 1994.

Roth, Nancy. *The Breath of God: An Approach to Prayer*. Cambridge, Massachusetts: Cowley Publications, 1990.

Roche de Coppens, Peter. "The Role of Prayer in Esoteric Christianity," in *Divine Light and Love*. Rockport, Massachusetts: Element Books, 1994.

Schneider, Marius. "On Gregorian Chant and the Human Voice," *World of Music*, XXIV/3, 1982.

Schneider, Marius, "Primitive Music," in Egon Wellesz, ed., *The New Oxford History of Music: Vol. I Ancient and Oriental Music*. London: Oxford University Press, 1957.

Snodgrass, Cynthia. "Musical Mythology," in Pratt and Spintge, eds., *MusicMedicine*, Vol. II. St. Louis: Magna Music Baton, 1996.

Spintge, Ralph. "The Anxiolytic Effects of Music," in Matthew H.M. Lee, Ed., Rehabilitation, Music and Well-Being, 82–97.

Steindl-Rast, David. *The Music of Silence*. San Francisco: Harper, 1994.

Chapter 7

Alper, Harvey P., Ed. *Understanding Mantras*. Albany, New York: SUNY Press, 1989.

Bush, Carol. *Healing Imagery and Music*. Portland, Oregon: Rudra Press, 1995.

Campbell, Don G. *Music: Physician for Times to Come*. Wheaton, IL: Quest, 1991.

Campbell, Don G. *The Roar of Silence: Healing Powers of Breath, Tone and Music*. Wheaton, IL: Quest, 1989.

Clifford, Patricia Hart. *Sitting Still*. New York: Paulist Press, 1994.

Crandall, Joanne. *Self-Transformation through Music*. Wheaton, Illinois: Theosophical Publishing, 1986.

Easwaran, Eknath. *The Unstruck Bell*. Tomales, California: Nilgiri Press, 1993.

Finley, Mitch. "Max Picard's Celebration of Silence." *Catholic Twin Circle*, June 4, 1995.

Franklin, Ursula. "Silence and the Notion of the Commons." *The Soundscape Newsletter*, January 1994.

Gawronski, Raymond. *Word and Silence*. Grand Rapids, Michigan: Eerdman's, 1995.

Hale, Susan Elizabeth. *Song and Silence*. Albuquerque: La Alameda Press, 1995.

Halpern, Steven and Louis Savary. *Sound Health*. San Francisco: Harper & Row, 1985.

Schafer, R. Murray. "Silences," in *The Soundscape: Our Sonic Environment and the Tuning of the World.* Rochester, Vermont: Destiny Books, 1994.

Singh, Sawan. *Philosophy of the Masters*. Vol. IV. Punjab, India: Radha Soami Satsang Beas, 1989.

Chapter 8

Campbell, Don G. *The Roar of Silence: Healing Powers of Breath, Tone and Music.* Wheaton, IL: Quest, 1989.

Campbell, Don G. Music: *Physician for Times to Come.* Wheaton, IL: Quest, 1991.

Eliade, Mircea. *Shamanism: Archaic Techniques of Ecstasy.* New York: Pantheon, 1964.

Filippi, Gian Guiseppe. *Mrtyu: Concept of Death in Indian Traditions.* New Delhi: Printworld Ltd., 1996.

Gardner, Kay. *Sounding the Inner Landscape.* Stonington, Maine: Caduceus Publications, 1990.

Goldman, Jonathan. *Healing Sounds*. Rockport, Massachusetts: Element Books, 1992.

Goodman, Felicitas. *Where the Spirits Ride the Wind.* Bloomington, Indiana: Indiana University Press, 1990.

Gore, Belinda. *Ecstatic Body Postures*. Santa Fe: Bear & Co., 1995.

Harner, Michael. *The Way of the Shaman*. San Francisco: Harper, 1980.

Hykes, David. "Harmonic Chant–Global Sacred Music" in Don Campbell, *Music: Physician for Times to Come*. Wheaton, Illinois: Quest Books, 1991.

Ingerman, Sandra. *Soul Retrieval*. San Francisco: Harper, 1991.

Kalweit, Holger. *Dreamtime and Inner Space: The World of the Shaman*. Boston: Shambhala, 1984.

Lazar, Dr. Katalin. Discussions at the Institute for Musicology in Budapest, Hungary, during the summer of 1994.

Maxfield, Melinda C. "The Journey of the Drum," in Don Campbell, ed., *Music and Miracles*. Wheaton, Illinois: Quest Books, 1992.

Rouget, Gilbert. *Music and Trance: A Theory of the Relations between Music and Possession*. Chicago: University of Chicago Press, 1985.

Schneider, Marius. "Acoustic Symbolism," in Joscelyn Godwin, ed., *Cosmic Music: Musical Keys to the Interpretation of Reality*. Rochester, Vermont: Inner Traditions, 1989.

Schneider, Marius, "Primitive Music," in Egon Wellesz, ed., *The New Oxford History of Music: Vol. I Ancient and Oriental Music*. London: Oxford University Press, 1957.

Vitebsky, Piers. *The Shaman*. Boston: Little, Brown & Co., 1995.

Walsh, Roger N. *The Spirit of Shamanism*. New York: G.P. Putnam's Sons, 1990.

Walker, Barbara. *The Woman's Dictionary of Symbols and Sacred Objects*. San Francisco: Harper, 1988.

Weeks, Bradford. "The Physician, the Ear and Sacred Music," in Don Campbell, *Music: Physician for Time to Come*. Wheaton, Illinois: Quest Books, 1991.

Chapter 9

Abram, David. *The Spell of the Sensuous: Perception and Language in a More-than-Human World*. New York: Pantheon Books, 1996.

Bierlien, J.F. *Parallel Myths*. New York: Ballantine Books, 1994.

Boissiere, Robert. *Meditations with the Hopi*. Norman, Oklahoma: University of Oklahoma Press, 1986.

Bruchac, Joseph. *The Native American Sweat Lodge*. Freedom, California: The Crossing Press, 1993.

Chatwin, Bruce. *The Songlines*. New York: Penguin, 1987.

Eaton, Evelyn. *I Send a Voice*. Wheaton, Illinois: Theosophical Publishing, 1990.

Gill, Sam and Irene Sullivan. *Dictionary of Native American Mythology*. Oxford: Oxford University Press, 1992.

Gill, Sam D. *The Songs of Life: An Introduction to Navajo Religious Culture*. Leiden: EJ Brill, 1979.

Johnson, Sandy. *The Book of Elders*. San Francisco: Harper, 1994.

Mails, Thomas E. *Secret Native American Pathways*. Tulsa: Council Oak Books, 1997.

Meadows, Kenneth. *The Medicine Way*. Rockport, Massachusetts: Element, 1997.

Nabokov, Peter, ed. *Native American Testimony*. New York: Viking Penguin, 1991.

Powers, William K. *Sacred Language: The Nature of Supernatural Discourse in Lakota*. Norman, Oklahoma: University of Oklahoma Press, 1986.

Radin, Paul, "Music and Medicine among Primitive Peoples," in Dorothy Schullian and Max Schoen, *Music and Medicine*. New York: Schuman, 1948.

Rael, Joseph. *Being and Vibration*. Tulsa: Council Oak Books, 1993.

Ross, A.C. *Mitakuye Oyasin—We Are All Related*. Kyle, South Dakota: BEAR, 1991.

Sandner, Donald. *Navajo Symbols of Healing*. Rochester, Vermont: Healing Arts Press, 1991.

Swann, Brian. *Song of the Sky: Versions of Native American Song Poems*. Amherst, Massachusetts: University of Massachusetts Press, 1993.

Chapter 10

Andrew, Ted. "The Renaissance of the Bardic Traditions," Part II in *Sacred Sounds*. St. Paul, Minnesota: Llewellyn Publications, 1992.

Cirlot, J.E. *A Dictionary of Symbols*. New York: Philosophical Library, 1962.

Gary, Richard, F.S.A. (Scot). Conversations concerning Celtic spirituality and tradition held in spring of 1995.

Gary, Richard, F.S.A. (Scot). "The Harp of the Boyne," privately published, May 1990.

Glassie, Henry. *Irish Folk Tales*. New York: Pantheon, 1985.

Godwin, Joscelyn. *Harmonies of Heaven and Earth*. Rochester, Vermont: Inner Traditions, 1987.

Hale, Susan Elizabeth. *Song and Silence: Voicing the Soul*. Albuquerque: La Alameda Press, 1995.

Jackson, Kenneth H. *A Celtic Miscellany*. London: Penguin, 1971.

MacCullouch, J.A. *The Religion of the Ancient Celts*. London: Constable, 1992.

Mag Fhearaigh, Criostoir and Tim Stampton. *Ogham*. Malin, County Donegal: Ballagh Studios, 1997.

Matthews, John. *The Celtic Shaman*. Rockport, Massachusetts: Element, 1991.

Schneider, Marius. "Acoustic Symbolism," in Joscelyn Godwin, ed., *Cosmic Music: Musical Keys to the Interpretation of Reality*. Rochester, Vermont: Inner Traditions, 1989.

Sheldrake, Philip. *Living between Worlds: Place and Journey in Celtic Spirituality*. Boston: Cowley, 1995.

Walker, Barbara. *The Woman's Dictionary of Symbols and Sacred Objects*. San Francisco: Harper, 1988.

Williams, Sarajane. "The Harp of Dagda," *Harp Therapy Journal*, Summer 1997.

Chapter 11

Abram, David. *The Spell of the Sensuous: Perception and Language in a More-than-Human World*. New York: Pantheon Books, 1996.

Bhaskharananda. *Essentials of Hinduism*. Seattle: Viveka Press, 1995.

Campbell, Don G. *The Roar of Silence: Healing Powers of Breath, Tone and Music*. Wheaton, IL: Quest, 1989.

Chatwin, Bruce. *The Songlines*. New York: Penguin, 1987.

Davis, Avram. *The Way of Flame*. San Francisco: Harper, 1996.

Davis, Avram and Manuela Dunn Mascetti. *Judaic Mysticism*. New York: Hyperion, 1997.

Ernst, Carl W. *The Shambhala Guide to Sufism*. Boston: Shambhala, 1997.

Gray, William G. *The Ladder of Lights*. York Beach, Maine: Samuel Weiser, 1981.

Helminski, Kabir. "Ecstasy and Sobriety," *Eye of the Heart*, II:1, 2–4.

Helminski, Kabir. "Spiritual Practice: Sustaining Presence through Zhikr," *Eye of the Heart*, (Mevlevi Order and Threshold Society). Summer 1997, 10.

Idel, Moshe. *Kabbalah: New Perspectives*. New Haven, Connecticut: Yale University Press, 1988.

Jaoudi, Maria. *Christian and Islamic Spirituality: Sharing a Journey*. Mahwah, New Jersey: Paulist Press, 1993.
Khan, Hazrat Inayat. *Music*. Claremont, California: Hunter House, 1988.

Khan, Hazrat Inayat. *The Development of Spiritual Healing*. Claremont, California: Hunter House, 1974.

Khan, Hazrat Inayat. *The Music of Life*. New Lebanon, New York: Omega Publications, 1983.

Khan, Hazrat Inayat. *The Mysticism of Music, Sound and Word. Vol. II: The Sufi Message*. Delhi: Motilal Banarsidass, 1994.

Khan, Hazrat Inayat. *The Sufi Message: 14 Volumes*. New Delhi: Shri Jainendra Press, 1990.

Kramer, Kenneth. *The Sacred Art of Dying: How World Religions Understand Death*. Mahwah, New Jersey: Paulist Press, 1988.

Mataji, Vandana. *Nama Japa: Prayer of the Name in the Hindu and Christian Traditions*. Delhi: Motilal Banarsidas, 1995.

Randhawa, G.S. *Guru Nanak's Japu Ji*. Amritsar: Guru Nanak Dev University, 1994.

Schneider, Marius, "Primitive Music," in Egon Wellesz, ed., *The New Oxford History of Music: Vol. I Ancient and Oriental Music*. London: Oxford University Press, 1957.

Scholem, Gershom. *On the Kabbalah and its Symbolism*. New York: Schocken Books, 1974.

Singh, Dewan. *Mysticism of Guru Nanak*. Amritsar: Singh Brothers, 1995.
Singh, Dewan. *Sikh Mysticism*. Amritsar: self-published, 1964.

Singh, Jodh. *The Religious Philosophy of Guru Nanak*. Delhi: B.K. Press, 1989.

Singh, Kirpal. *Naam or Word*. Bowling Green, Virginia: Sawan Kirpal Publications, 1981.
Singh, Sawan. *Philosophy of the Masters, Series I-IV*. Punjab: Radha Soami Satsang Beas, 1990.

Steinsaltz, Adin. *In the Beginning*. Northvale, New Jersey: Jason Aronson, 1995.

Steinsaltz, Adin. *The Sustaining Utterance*. Northvale, New Jersey: Jason Aronson, 1996.

Zaehner, R.C. *Hindu and Muslim Mysticism*. Oxford: Oneworld Publications, 1994.

Chapter 12

Abram, David. *The Spell of the Sensuous: Perception and Language in a More-than-Human World*. New York: Pantheon Books, 1996.

Alper, Harvey P., ed. *Understanding Mantras*. Albany, New York: SUNY Press, 1989.

Bhaskharananda. *Essentials of Hinduism*. Seattle: Viveka Press, 1995.

Chatterji, J.C. *The Wisdom of the Vedas*. Wheaton, Illinois: Theosophical Publishing, rev. ed.,1992.

Coward, Harold. "The Reflective Word: Spirituality in the Grammarian Tradition of India" in Krishna Sivaraman, ed. *Hindu Spirituality*. Delhi: Motilal Banarsidass, 1995.

Danielou, Alain. *Music and the Power of Sound: The Influence of Tuning and Interval on Consciousness*. Rochester, Vermont: Inner Traditions, repr. 1995.

Davis, Avram and Manuela Dunn Mascetti. *Judaic Mysticism*. New York: Hyperion, 1997.

Deshpande, V.W. *The Impact of Ancient Indian Thought on Christianity*. New Delhi: APH Publishing, 1996.

Dey, Suresh Chandra. *The Quest for Music Divine*. New Delhi: Ashish, 1990.

Dongre, Archana. "Fixing History," *Hinduism Today*, May 1998.

Elizarankova, Tatyana J. *Language and Style of the Vedic Rishis*. Albany, New York: SUNY Press, 1995.

Feuerstein, Georg. *The Yoga-Sutra of Patanjali*. Rochester, Vermont: Inner Traditions, 1989.

Feuerstein, Georg, Subhash Kak and David Frawley. *In Search of the Cradle of Civilization*. Wheaton, Illinois: Quest Books, 1995.

Frawley, David. "Sanskrit, the Language of the Vedas," in Vyaas Houston, ed., *Devavani*. Warwick, New York: American Sanskrit Institute, n.d.

Harding, Elizabeth U. *Kali: The Black Goddess of Dakshineswar*. York Beach, Maine: Nicolas-Hays, 1993.

Houston, Vyaas, ed. *Devavani: A Collection of Essays, Articles, and Quotes on Sanskrit*. Warwick, New York: American Sanskrit Institute, n.d.

Houston, Vyaas. *The Yoga Sutra Workbook*. Warwick, New York: American Sanskrit Institute, 1995.

Jansen, Eva Rudy. *The Book of Hindu Imagery*. Diever, Holland: Binkley Kok, 1997.

Johari, Harish. *Tools for Tantra*. Rochester, Vermont: Destiny Books, 1986.

Lawlor, Robert. "The Resounding Cosmos and the Myth of Desire," *Parabola*, Vol. 2.

LeMee, Jean. *Hymns of the Rig-Veda*. New York: Knopf, 1975.

Mircea, Eliade. *Yoga: Immortality and Freedom*. Princeton: Princeton University Press, 1990.

Padoux, Andre. *Vac: The Concept of the Word in Selected Hindu Tantras*. Albany, New York: SUNY Press, 1990.

Patnaik, Tandra. *Shabda: A Study of Bhartrhari's Philosophy of Language*. New Delhi: D.K. Printworld, 1994.

Schneider, Marius, "The Nature of the Praise Song," in Joscelyn Godwin, ed. *Cosmic Music*. Rochester, Vermont: Inner Traditions, 1989.

Schneider, Marius, "Primitive Music," in Egon Wellesz, ed., *The New Oxford History of Music: Vol. I Ancient and Oriental Music.* London: Oxford University Press, 1957.

Shannahoff-Khalsa, David and Yogi Bhajan. "The Healing Power of Sound: Techniques from Yogic Medicine," in Spingte and Droh, eds., *MusicMedicine.* St. Louis: Magna Music Baton, 1992.

Thite, G.U. *Music in the Vedas: Its Magico-Religious Significance.* New Delhi: Sharada, 1997.

Tyberg, Judith. *The Language of the Gods.* Los Angeles: East-West Cultural Center, 1967.

Walker, Barbara. *The Woman's Dictionary of Symbols and Sacred Objects.* San Francisco: Harper, 1988.

Woodroffe, Sir John. *The Garland of Letters.* Madras: Ganesh & Co., 1994.

Yogananda, Paramahansa. *Autobiography of a Yogi.* Bombay: Kothari, repr. 1994.

Chapter 13

Alper, Harvey P., ed. *Understanding Mantras.* Albany, New York: SUNY Press, 1989.

Bruyere, Rosalyn L. *Wheels of Light.* New York: Simon & Schuster, 1994.

Campbell, Don G. *The Mozart Effect: Tapping the Power of Music to Heal the Body, Strengthen the Mind, and Unlock the Spirit.* New York: Avon Books, 1997.

Campbell, Don G. *Music and Miracles.* Wheaton, IL: Quest, 1992.

Campbell, Don G. *The Roar of Silence: Healing Powers of Breath, Tone and Music.* Wheaton, IL: Quest, 1989.

Chawdhri, L.R. *Practicals of Mantras and Tantras.* New Delhi: Sagar, 1995.

Cornell, Judith. *Mandala.* Wheaton, Illinois: Theosophical Publishing, 1994.

Coward, Harold and David Goa. *Mantra: Hearing the Divine in India.* Chambersburg, Pennsylvania: Anima, 1991.

Danielou, Alain. *Music and the Power of Sound: The Influence of Tuning and Interval on Consciousness.* Rochester, Vermont: Inner Traditions, repr. 1995.

Dange, Sindhu S. *Aspects of Speech in Vedic Ritual.* New Delhi: Aryan Books, 1996.

Dyczkowski, Mark S.G. *The Doctrine of Vibration: An Analysis of the Doctrines and Practices of Kashmir Shaivism.* Albany, New York: SUNY Press, 1987.

Easwaran, Eknath. *The Unstruck Bell.* Tomales, California: Nilgiri Press, 1993.

Elizarenkova, Tatyana J. *Language and Style of the Vedic Rishis.* Albany, New York: SUNY Press, 1995.

Evola, Julius. *The Yoga of Power.* Rochester, Vermont: Inner Traditions, 1992.

Feuerstein, Georg. "The Sacred Syllable OM," Yoga Research Center Studies Series, No. 2, 1997.

Feuerstein, Georg, Subhash Kak and David Frawley. *In Search of the Cradle of Civilization.* Wheaton, Illinois: Quest Books, 1995.

Holtje, Dennis. *From Light to Sound.* Albuquerque: Master Path, 1995.

Hughes, John. *Self-Realization in Kashmir Shaivism: The Oral Teachings of Swami Lakshmanjoo*. Albany, New York: SUNY Press, 1994.

Johari, Harish. *Chakras: Energy Centers of Transformation*. Rochester, Vermont: Destiny Books, 1987.

Johari, Harish. Chapter 3 in *Tools for Tantra*. Rochester, Vermont: Destiny Books, 1986.

Juergensmeyer, Mark. *Radhasoami Reality: The Logic of a Modern Faith*. Princeton: Princeton University Press, 1991.

Khanna, Madhu. "Archetypal Space and Sacred Sound," in *Yantra*. London: Thames and Hudson, 1997.

Lawlor, Robert. "The Resounding Cosmos and the Myth of Desire," *Parabola*, Vol. 2.

LeMee, Jean. *Hymns of the Rig-Veda*. New York: Knopf, 1975.

McClellan, Randall. *The Healing Forces of Music: History, Theory, and Practice*. Rockport, Massachusetts: Element Books, 1991.

Padoux, Andre. *Vac: The Concept of the Word in Selected Hindu Tantras*. Albany, New York: SUNY Press, 1990.

Rael, Joseph. *Being and Vibration*. Tulsa: Council Oak Books, 1993.

Roche de Coppens, Peter. *Divine Light and Fire*. Rockport, Massachusetts: Element Books, 1992.

Roche de Coppens, Peter. *Divine Light and Love*. Rockport, Massachusetts: Element Books, 1994.

Rudhyar, Dane. *The Magic of Tone and the Art of Music*. London: Shambhala, 1982.

Sant Keshavadas. *Gayatri: The Highest Meditation*. Delhi: Motilal Banarsidass, 1994.

Schneider, Marius. "Acoustic Symbolism," in Joscelyn Godwin, ed. *Cosmic Music: Musical Keys to the Interpretation of Reality*. Rochester, Vermont: Inner Traditions, 1989.

Schneider, Marius. "The Nature of the Praise Song," in Joscelyn Godwin, ed. *Cosmic Music*. Rochester, Vermont: Inner Traditions, 1989.

Schneider, Marius. "Primitive Music," in Egon Wellesz, ed., *The New Oxford History of Music: Vol. I Ancient and Oriental Music*. London: Oxford University Press, 1957.

Sharma, Shri Ram. *The Great Science and Philosophy of Gayatri*. Mathura, India: Yug Nirman Yojna, n.d.

Shubhakaran, K. T. *Mystical Formulae*. New Delhi: Sagar, 1996.

Sivananda. *Japa Yoga*. Dt. Tehri-Garhwal, U.P., Himalayas, India: The Divine Life Society, 1972.

Staal, Frits. *Ritual and Mantras: Rules without Meaning*. Delhi: Motilal Banarsidass, 1996.

Stewart, R.J. *The Spiritual Dimension of Music*. Rochester, Vermont: Destiny, 1987.

Tigunat, Pandi Rajmani. "The Bridge to the Inner World," in *The Power of Mantra and the Mystery of Initiation*. Honesdale, Pennsylvania: Himalayan International Institute of Yoga, 1996.

Twitchell, Paul. *The Shariyat-Ki-Sugmad, Book II*. Crystal, Minnesota: Illuminated Way, 1971.

Yogananda, Paramahansa. *Autobiography of a Yogi*. Bombay: Kothari, repr. 1994.

Epilogue

Abhishiktananda. *Hindu-Christian Meeting Point.* Delhi: ISPCK, 1983.

Berendt, Joachim-Ernst. *The Third Ear: On Listening to the World.* New York: Holt and Co., 1985.

Berendt, Joachim-Ernst. *The World Is Sound: Nada Brahma—Music and the Landscape of Consciousness.* Rochester, Vermont: Destiny Books, 1987.

Campbell, Don G. *The Mozart Effect: Tapping the Power of Music to Heal the Body, Strengthen the Mind, and Unlock the Spirit.* New York: Avon Books, 1997.

Campbell, Don G. *The Roar of Silence: Healing Powers of Breath, Tone and Music.* Wheaton, IL: Quest, 1989.

Crandall, Joanne. *Self-Transformation through Music.* Wheaton, Illinois: Theosophical Publishing, 1988.

Danielou, Alain. *Music and the Power of Sound: The Influence of Tuning and Interval on Consciousness.* Rochester, Vermont: Inner Traditions, repr. 1995.

Davis, Avram and Manuela Dunn Mascetti. *Judaic Mysticism.* New York: Hyperion, 1997.

Easwaran, Eknath. *The Unstruck Bell.* Tomales, California: Nilgiri Press, 1993.

Hale, Susan Elizabeth. *Song and Silence: Voicing the Soul.* Albuquerque: La Alameda Press, 1995.

Halpern, Steven and Louis Savary. *Sound Health: The Music and Sounds that Make Us Whole.* San Francisco: Harper & Row, 1985.

Hamel, Peter Michael. *Through Music to the Self.* Longmead, England: Element Books, 1991.

Heline, Corinne. *Color and Music in the New Age.* Marina del Rey, California: De Vorss, 1987.

Heline, Corinne. *Esoteric Music.* Marina del Rey, California: DeVorss, 1969.

Heline, Corinne. *Music: The Keynote of Human Evolution.* Santa Monica, California: New Age Bible and Philosophy Center, 1986.

Hubbard, Barbara Marx. "Music for the Birth of Planet Earth" in Don Campbell, ed., *Music and Miracles.* Wheaton, IL: Quest Books, 1992.

Khan, Hazrat Inayat. *The Mysticism of Sound.* New York: Weber, 1979.

Kalliath, Anthony. *The Word in the Cave.* New Delhi: Intercultural Publications, 1996.

Khan, Hazrat Inayat. *The Music of Life.* New Lebanon, New York: Omega Publications, 1983.

Khan, Hazrat Inayat. *The Mysticism of Music, Sound and Word. Vol II: The Sufi Message.* Delhi: Motilal Banarsidass, 1994.

Kramer, Kenneth. *The Sacred Art of Dying: How World Religions Understand Death.* Mahwah, New Jersey: Paulist Press, 1988.

Leonard, George. *The Silent Pulse.* New York: Bantam, 1981.

McClellan, Randall. *The Healing Forces of Music: History, Theory, and Practice.* Rockport, Massachusetts: Element Books, 1991.

Main, John. *Word into Silence.* New York: Paulist, 1981.

Maclean, Norman. *A River Runs Through It*. Chicago: University of Chicago Press, 1976.

Schafer, R. Murray. *The Book of Noise*. Wellington, New Zealand: Price Milburn Co., 1973.

Schafer, R. Murray. *The Soundscape: Our Environment and the Tuning of the World*. Rochester, Vermont: Destiny Books, 1994.

Singh, Dewan. *Sikh Mysticism*. Amritsar: self-published, 1964.

Stewart, R.J. *The Spiritual Dimension of Music*. Rochester, Vermont: Destiny, 1987.

Tame, David. *The Secret Power of Music: The Transformation of Self and Society through Musical Energy*. Rochester, Vermont: Destiny Books, 1984.

Acknowledgments

When acknowledging my gratitude and indebtedness for a work so broad in scope, and one which has touched and shaped so much of my life, I am acutely aware of just how many people I need to thank or honor. It is my prayer that the naming and acknowledgement herein will be complete, but I also include a deep apology to anyone unnamed accidentally due to the last-minute rush of publishing I begin with the obvious by gratefully thanking my parents for so many fundamental life supports and formational influences–my father for his love of language, poetry and song, and for his sharing them with me, and my mother (who raised me largely as a single parent), who sacrificed so much so I could experience and participate in the arts at a young age. I also wish to thank John Kohrs, who gave me my first musical instrument, and his wonderfully warm family, who helped shape my early love of music.

This work would never have come into being without the financial support of the Fetzer Institute in Kalamazoo, Michigan, and the very exceptional emotional and spiritual support of Robert F. Lehman and Carol Hegedus, who were intrigued as to how sound might effect the human spirit and who trustingly encouraged the finding of a sonic vision not yet imagined, heard, or seen.

I would like to thank Vyaas Houston, founder of the American Sanskrit Institute, for the power of his teaching, the depth of his spirit in creating sacred space, and his love of the intricate mysteries found within the sounds of the Sanskrit language which have opened so much beauty for me, created so much of the inspiration and impetus for this piece, and through which I have experienced the altar of sound.

I am grateful for the guidance of many teachers. Primarily, I am indebted to: the late Dr. George Nugent of

Syracuse Univeristy, for his exacting but always supportive musicological training; John Monkman, a divine voice teacher who, through the knowledge of *bel canto* technique, first taught me to spin threads of sacred sound, and Dr. John Hsu of Cornell University, virtuoso cellist and viola da gamba player, who so patiently trained me to "cry with one eye."

For conversations early on, I would like to thank musicians on the growing edge of music and healing: Don Campbell, ground-breaking pioneer of sound and healing; Pat Moffitt Cook, modern-day herald of the ancient practices of indigenous sound healing; and Therese Schroeder-Sheker, devoted advocate for music at the bedside. I am deeply indebted to Sue Richards, a magically-inspired harper, for her knot-releasing inspiration on the Celtic harp and to Richard Gary for his guidance concerning Celtic spiritual traditions. I am deeply indebted to Leroy Little Bear, Amethyst First Rider, Tobasonakwut Kinew, Henry and LouAnn Bush for powerful dialogues concerning Native American traditions of sacred sound and practice. For an introduction to Sufi spiritual practices of sound and worship, I am grateful for conversations with and practices attended by Shaikh Kabir Helminski of the Mevlevi Order and the Threshold Society.

I give thanks for the spiritual depths of the Episcopal Church, which stands so firmly secure in the Word that through its broad traditions it remains open to and supportive of the truths found in other spiritual traditions. I am grateful for the many gifts given to me over the past twenty years by the hundreds of hospice patients whom I have served and to whom I have had the great honor of singing sacred songs. I am grateful for the training and guidance of Rev. Terry Ruth Culbertson, my mentor in hospice and general ministry. She, Dr. Sherry Magill, and Susan Dorn, Esq., have provided invaluable support, advice, and nurturance for the creation of the Sacred Sound Institute.

I give thanks to those at Paraview Press, especially Sandra Martin, Claire Wyckoff, and Erika Lieberman, for their support, encouragement, and dedication to getting this volume in print.

I give continual thanks to my husband, Patrick, who has been so supportive, so grounding, and so understanding while I journeyed from one place to the other (both externally and internally) collecting impressions and information for this piece. The counterpoint of our marriage continually brings such light, laughter, and harmony to my life.

Finally, I am eternally grateful for the great goodness of God who has sent all these wonderful teachers, guides, and gurus into my life.

Want to know more about Paraview Press books?

Descriptions of the following titles and ordering information are available at www.paraview.com. Paraview titles are immediately available on amazon.com and other online bookstores. They also may be special-ordered through your local bookstore.

Any Woman Can!
Sheila Grant

Dancing With the Wind:
A True Story of Zen in the Art of Windsurfing
Laurie Nadel

Dream Interpretation (and more!) Made Easy
Kevin Todeschi

Fatal Attractions: The Troubles with Science
Henry H. Bauer

Flowers that Heal: Aromas, Herbs, Essences, and Other Secrets of the Fairies
Judy Griffin

In the Big Thicket: On the Trail of the Wild Man
Rob Riggs

In Realms Beyond: Book One of the Peter Project
Al Miner and Lama Sing

Misdiagnosed: Was My Wife a Casualty of America's Medical Cold War?
A. Robert Smith

Mothman and Other Curious Encounters
Loren Coleman

Mysterious America: The Revised Edition
Loren Coleman

The Perfect Horoscope
John Willner

River of a Thousand Tales: Encounters with Spirit, Reflections from Science
Rao Kolluru

The Siren Call of Hungry Ghosts
Joe Fisher

Spiritual Places In and Around New York City
Emily Squires and Len Belzer

Swamp Gas Times: My Two Decades on the UFO beat
Patrick Huyghe

Trauma Room One
Charles A. Crenshaw, M.D.

Paraview Press uses digital print-on-demand technology (POD), a revolution in publishing which makes it possible to produce books without the massive printing, shipping and warehousing costs that traditional publishers incur. In this ecologically friendly printing method, books are stored as digital files and printed one copy at a time–or more–as demand requires. Now high-quality paperback books can reach you, the reader, faster than ever before. We believe that POD publishing empowers authors and readers alike. Free from the financial limitations of traditional publishing, we specialize in topics for niche audiences such as Mind/Body/Spirit and Frontiers of Science and Culture. Please visit our website for more information.

Paraview Press

Printed in the United States
131957LV00003B/123/A